Great Scientific Experiments

Frontispiece: *L'Académie des Sciences et des Beaux Arts*, engraving by C. N. Cochin (after Le Clerc), 1698.

N.C. Sculp.

L'ACADEMIE DES SCIENCES
DEDIÉE
Par son tres humble tres obéissant et tres

ET DES BEAUX ARTS
AU ROY.
fidele Serviteur et sujet Seb. le Clerc.

Great
Scientific
Experiments

20 Experiments that
Changed our View of the World

Rom Harré

Phaidon

OXFORD

1981

Phaidon Press Limited, Littlegate House, St Ebbe's Street, Oxford

First published 1981
© 1981 by Phaidon Press Limited
All rights reserved

British Library Cataloguing in Publication Data

  Harré, Romano
    Great Scientific experiments
    1. Science – Experiments – History
    I. Title
    509'.2'2   Q125

    ISBN 0-7148-2096-2

Phototypeset by Tradespools Ltd, Frome, Somerset
Printed and bound in Great Britain by Morrison and Gibb Limited, Edinburgh

Acknowledgements

Frontispiece: Trustees of the British Museum. Fig. 1: Cavendish Laboratory, Cambridge (photo C. E. Wynn-Williams). Figs 3, 18, 20, 21, 22, 23, 24, 25, 26, 27, 40, 42, 43, 44, 46, 65, 71, 74, 89, 90, 92, 93, 95, 105, 108, 110, 111: Museum of the History of Science, Oxford University. Figs 5, 6: Mary Evans Picture Library, London. Fig. 8: Deutsches Archäologisches Institut, Rome. Fig. 9: Zurich, Zentralbibliothek. Figs 10, 13, 14, 17, 19, 36, 41, 47, 48, 55, 78, 97, 101, 107, 127, 128: Bodleian Library, Oxford. Figs 11, 12: National Library of Medicine, Bethesda. Figs 15, 16: President and Council of the Royal College of Surgeons of England (Medical Illustration Support Service). Figs 28, 30, 32: Hermann Kacher, Seewiesen. Figs 29, 96: BBC Hulton Picture Library. Fig. 31: Methuen Ltd (Konrad Lorenz, *King Solomon's Ring*, 1952). Fig. 33: J. M. Dent Ltd (A. Nisbett, *Konrad Lorenz*, 1976). Fig. 34: Musées Nationaux, Paris. Fig. 37: Alinari, Florence. Fig. 38: Istituto e Museo di Storia della Scienza, Florence. Fig. 39: Whipple Museum of the History of Science, Cambridge (photo Edward Leigh). Figs 49, 99: Trustees of the British Library. Figs 50, 51, 53: Universitäts-Bibliothek, Basel. Fig. 54: Universitäts-Bibliothek, Erlangen. Figs 56, 58, 62, 104: Mansell Collection, London. Fig. 57: Wellcome Trustees, Wellcome Institute for the History of Medicine. Fig. 59: Musée Pasteur, Paris. Figs 60, 67, 100, 120: Cavendish Laboratory, Cambridge. Fig. 61: From the collection of Dr Alfred Bader (photo Ken Brown). Figs 64, 91, 103, 117: Science Museum, London. Fig. 68: CERN. Figs 69, 70: Case Western Reserve University. Fig. 77: Mount Wilson and Las Campanas Observatories, Pasadena, California. Fig. 79: Institut Pasteur, Paris. Fig. 81: J. D. Watson © 1968. Figs 80, 82, 83, 84: Blackwell Scientific Publications. Fig. 85: Mrs E. Gibson, Ithaca, New York. Fig. 88: Metropolitan Museum of Art, New York. Fig. 94: Eidgenössische Technische Hochschule, Zurich. Figs 113, 114: Royal Institution, London. Fig. 115: Illustrated London News Picture Library. Figs 116, 118: Royal Swedish Academy of Sciences, Stockholm. Fig. 121: Nobel Foundation, Stockholm. Fig. 119: Deutsches Museum, Munich. Fig. 122: Professor Dr Walther Gerlach.
Figs 2, 4, 7, 35, 45, 52, 63, 66, 72, 73, 75, 76, 82, 86, 87, 98, 102, 106, 112, 123, 124, 125, 126 were drawn for this book by Illustration Services, Oxford.

Contents

Preface   7

Introduction   9

I. Formal Aspects of Method

A. Exploring the Characteristics of a Naturally Occurring Process
   1. Aristotle: *The Embryology of the Chick*   31
   2. William Beaumont: *The Process of Digestion as Chemistry*   39

B. Deciding between Rival Hypotheses
   3. Robert Norman: *The Discovery of Dip and the Field Concept*   49
   4. Stephen Hales: *The Circulation of Sap in Plants*   57
   5. Konrad Lorenz: *The Conditions of Imprinting*   65

C. Finding the Form of a Law Inductively
   6. Galileo: *The Law of Descent*   76
   7. Robert Boyle: *The Measurement of the Spring of the Air*   83

D. The Use of Models to Simulate otherwise Unresearchable Processes
   8. Theodoric of Freibourg: *The Causes of the Rainbow*   93

E. Exploiting an Accident
   9. Louis Pasteur: *The Preparation of Artificial Vaccines*   102
   10. Ernest Rutherford: *The Artificial Transmutation of the Elements*   111

F. Null Results
   11. A. A. Michelson and E. W. Morley: *The Impossibility of Detecting the Motion of the Earth*   124

II. Developing the Content of a Theory

A. Finding the Hidden Mechanism of a Known Effect
   12. F. Jacob and E. Wollman: *The Direct Transfer of Genetic Material*   137
   13. J. J. Gibson: *The Mechanism of Perception*   147

6

Contents

B. Existence Proofs
14. A. L. Lavoisier: *The Proof of the Oxygen Hypothesis*   155
15. Humphry Davy: *The Electrolytic Isolation of New Elements*   163
16. J. J. Thomson: *The Discovery of the Electron*   171

C. The Decomposition of an Apparently Simple Phenomenon
17. Isaac Newton: *The Nature of Colours*   182

D. The Demonstration of Underlying Unity within Apparent Variety
18. Michael Faraday: *The Identity of All Forms of Electricity*   191

III. Technique

A. Accuracy and Care in Manipulation
19. J. J. Berzelius: *The Perfection of Chemical Measurement*   201

B. The Power and Versatility of Apparatus
20. Otto Stern: *The Wave Aspect of Matter and the Third Quantum Number*   212

General Bibliography   221

Index   223

Preface

The idea for a collection of brief studies of great experiments came from the editorial staff of the publisher of this volume. To realize such a project in a manageable form some compromises and adaptations had to be made. In its final shape I have planned this book not only to tell twenty stories but also to show the diverse roles that experiments play in science.

It is not possible to explain the significance of experiments drawn from many fields and many historical periods without making some assumptions about the scientific background of one's potential readers. While I have tried to make everything as clear as possible I have thought of myself as writing for someone who has had some acquaintance with the natural sciences. I have kept in mind a reader who has at some time done a General Science course at school. Historical and philosophical studies of science should not only relate experiments to theories, but also to the social and cultural background within which they were conceived. Social influences, such as the economic demands of an epoch, not only direct the interest of the scientific community to one class of problems rather than another, but they have some influence too on the images of the world that lie at the foundations of theories. Some social historians of science have argued that such 'external' factors may even influence the very criteria by which experiments are judged successful and unsuccessful and theories true or false.

While common sense must support the idea that there are a host of influences between a society and its science, it has proved very difficult to trace these influences in concrete form. The task is formidable. One has not only to find a way of expressing the central themes of a period, but to develop plausible social psychological hypotheses about the relation between these themes, their unfolding to the active minds of a period, and the process of creation itself. So far no one has succeeded in bringing off a really plausible study of concrete scientific work in its specific social setting to show the influences at work. Each experiment described in this book would need its own treatise to relate it to the social conditions of the times in which it seemed good to its performer to carry it out. Having a philosophical rather than historical interest in experiments, I have expounded each experiment in relation only to its strictly scientific context, knowing that a full understanding of it would require very much more.

To strike the right level of accuracy of description with general intelligibility I have been greatly helped by Mr Bernard Dod, Dr I. J. R. Aitchison and Dr B. Cox. I am most grateful for their criticism and help. I am grateful to Oxford University Press for permission to reprint a passage from Aristotle's *Historia Animalium*. The illustrations have been selected by Dr W. Hackmann of the Museum of the History of Science, Oxford.

Linacre College, Oxford, July 1980

Introduction

The fascination of experiments is many-sided. The equipment itself has a special charm, an irresistible combination of gadgetry and work of art. I remember very well the satisfaction I took in the very physical presence of the apparatus in my first chemistry and electricity 'sets'. Then there are the sudden glimpses of a mysterious reality that come when the equipment is put to use. I vividly recall the night my father and I prepared bromine. I was nine years old and so the anticipated length of the experiment had called for some preliminary negotiations about bed time. The apparatus was set up on the kitchen table, and the heat from the spirit lamp gently applied. Suddenly a reddish-brown liquid began to condense in the stem of the retort. Here was something drawn from within the unpromisingly pale ingredients with which we had begun. Then from the successful experiment comes a special feeling of power. This feeling seems to me to give a modern person an insight into the alchemical and magical tradition from which experimental science partly originated. There is something enormously thrilling about getting experimental apparatus to work. When a galvanometer registers a current or the flocculent white precipitate congeals out of the liquid, for a moment one has a sense of the forces of nature subdued to one's will. This is the romantic side of experimentation. I think I detect a measure of fellow feeling between myself as a schoolboy, and the long line of experimentalists who felt their activities in some kind of cosmic frame. This feeling is apparent in the Alexandrian treatises that have come down to us from the first few centuries of the Christian era as the works of the mythical scientist Hermes Trismegistus. It is just as evident in the attitudes Michael Faraday expressed, when for a moment he allowed his deep convictions to show through. But the same feeling is the source of the disappointment that many university students feel as the tedium of second-year chemistry practicals begins to wear them down. How does this come about?

Experiments have other uses than to offer glimpses of a mysterious reality to the romantically inclined. They are the basis of tightly disciplined means for the acquisition of certified practical knowledge. Though the impulse to 'unlock the secrets of nature' may be romantic, the uses of those secrets can be quite utilitarian. In the end the 'troops of effects' that Bacon foresaw coming from illuminating experiments that reveal the 'latent process and configuration' behind the surface appearances of nature are the point of science for most people nowadays. But this was not always so. One might be forgiven for thinking that the role of experiments in the production of certified knowledge could not be more obvious. In phrases like 'unlocking the secrets of nature' there seems to be embodied an image like that of Pandora's box. If you want to know what is in the box simply open the lid and have a look. The consequences may be problematic but the image suggests that the method of inquiry is not the difficulty. But it is not so simple. The lid of the box is usually obstinately stuck fast. All one has to go on are the strange noises that sometimes can be heard in response to one's knocking. And even when one does prise open the lid a crack, how does one recognize what one glimpses within? Without some prior idea of what to expect, the results of experimental science are usually opaque. Because the matter is so complex there has been room for very different views as to the role of experiments in science, each emphasizing an aspect of the systematic questioning of nature.

Looked at from these different points of view, experiments will be seen to have very different force. I hope to show, in this introduction, that rather than being rivals, the various theories of the experiment can be fitted together into a comprehensive understanding of the empirical side of the process of scientific discovery. We will need this comprehensive understanding to appreciate fully the experiments to be described and illustrated in what is to come.

The criteria for choosing the experiments described in this book

I suppose that in all hundreds of thousands, perhaps millions of experiments have been done since the Greeks began systematic scientific studies about 400 years before Christ. To find twenty that would serve both to entertain and instruct, some pretty strong criteria were needed.

There are experiments that are so well known, or at least have been heard of so widely, that they choose themselves. However, their very fame and the fact that they are described so often in textbooks and classrooms have slowly distorted the story of some of them, and the common image is sometimes quite inaccurate. For that reason I have used no secondary sources for the research for this book. Each experiment is described on the basis of the original paper or book in which the result was first announced. Two famous experiments that have become distorted in popular consciousness are those of Michelson and Morley, and Boyle. The Michelson–Morley experiment is widely but erroneously believed to have been the source of Einstein's Special Theory of Relativity. The discovery of Boyle's Law was not motivated by a disinterested curiosity about the physical properties of gases, but was meant as a knock-out blow against those theologian–physicists who denied the possibility of the vacuum. Pasteur is widely and correctly believed to have discovered the method for creating artificial vaccines. But how widely is it known that it all came about through his taking an extended summer holiday?

Fame is not always the best index of historical importance. The criterion of historical importance is itself somewhat equivocal, since the things that seem to us to have been important are highlighted by hindsight. I have tried to pick out experiments that were influential in their own times, as far as I can guess, and which have continued to reverberate through the subsequent development of a field of study. Theodoric's masterly investigation of the causes of the rainbow is known to have influenced his successors directly, and to have had a permanent effect in popularizing the use of geometry in physics. Aristotle's study of the embryology of the chick can be traced with some certainty as the seminal work from which all embryological studies, including those of our own day, have been derived. Newton's optical experiments not only established a certain theory of colour on a firm foundation but provided an exemplar of systematic scientific work that was widely admired and copied. Hales's pioneering study of the physiology of plants must be included in this category for another reason. Not only did he solve the outstanding problems suggested by the anatomical and theoretical work of Grew and Harvey, but he demonstrated that a certain kind of life process, namely the hydrostatics and hydrodynamics of the fluids in living beings, can be studied experimentally. I have chosen to illustrate his work with a single experiment on the circulation of plants, but his greatest triumphs were in the investigation of animal circulatory systems, confirming what Harvey had but guessed about the plumbing of the mammalian body.

My third criterion was more aesthetic. I have tried to select some experiments for their elegance, neatness and style. With the slightest of means an experimenter of genius goes right to the heart of a problem and transforms our understanding. Norman's simple wine-glass experiment, with which he and Gilbert were satisfied they had demonstrated the existence of a magnetic field (and not just magnetic attractions), has this quality. It must retain its place, even though later generations of scientists were able to show that with more sophisticated mathematics magnetic phenomena could, after all, be explained by forces of attraction and repulsion. Norman's experiment exerted its seminal influence on subsequent thought through an inspired misinterpretation of the effect. But the acme of such experiments must surely be J. J. Gibson's 'cookie-cutter' experiment. The very foundations of the traditional psychology of perception were overturned with the help of a few items of kitchen equipment.

There are some serious misapprehensions as to how experiments give us knowledge. My fourth criterion was slanted to more practical matters. I wanted to dispel the idea that experiments are isolated events that stand by themselves. Most experiments are steps in a sequence of studies through which a vaguely delineated subject-matter is explored. Sometimes the experiment I have picked out to illustrate the kind of investigations that are typical of a programme might be thought to be a culmination or turning point in the research, but that is usually a judgement of hindsight. The importance of sustained explorations of a field is so great in the history of science that I have shown the man who was perhaps the greatest of all experimental scientists, Michael Faraday, at work on a painstaking systematic study, made up of many little experiments. Each successful demonstration adds to the weight of the important conclusion that there is really only one kind of electricity. In similar ways Rutherford's discovery of artificial transmutation of the elements and Thomson's successful measurement of the physical properties of subatomic particles, seminal though they may seem to us, were for the experimenters themselves steps in a programme.

I have tried to illustrate this by showing how the experiment that serves as a focus for the story of each section is part of a process. Most experiments are part of programmes which already have a history when the experiment is performed, and they contribute to the future of the programme by suggesting new lines of research and helping to close off others. As a research programme goes on, past experiments quite often come to be differently interpreted from how they were understood when they were first performed. Lavoisier thought he had discovered not only the physical basis of combustion, but the principle of acidity. For some time the word 'oxygen' ('acid-producer') was taken literally. It remained for Davy to show that some acids did not contain oxygen, and for Lavoisier's discovery to take on a different complexion.

Theories of the experiment

Why do scientists do experiments? The answer seems as obvious as the question seems banal: to find out about nature.

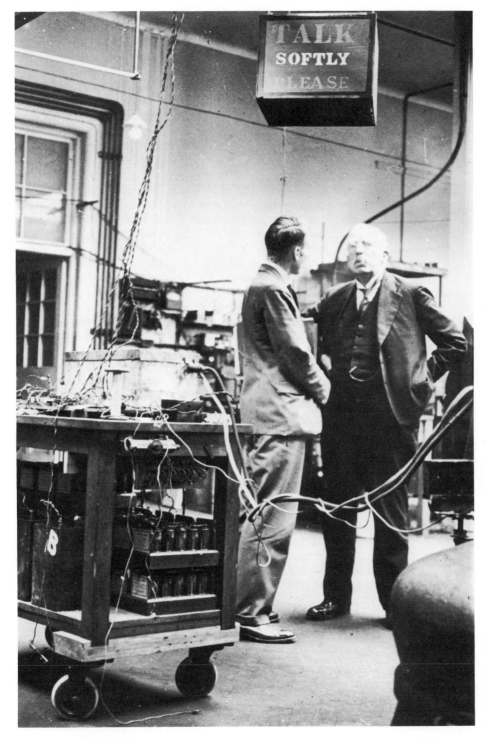

Fig. 1: Lord
Rutherford and
J. A. Ratcliffe in
conversation in
the Cavendish
Laboratory at
Cambridge. See
pp. 111–22.

But how do we formulate the most telling questions to put to nature, and how do we grasp what seem to be the answers? As we shall see, the world of ideas is very much mixed up with the world of facts. Without some prior idea of what might be there to be found out we would not know what to look for in the results of our experiments, nor would we be able to recognize it when we had found it. The point is vividly illustrated by the way accidents and chance events prompt discoveries. Only a mind prepared to recognize the significance of what has happened accidentally can draw a discovery out of it. Experimentalists of genius, like Faraday, generally knew exactly what to expect from their experiments, so powerful were their theories. These are the experimentalists who keep on nagging away until the experiment 'works'. When Pasteur discovered artificial vaccines an accidental event was significant to him, and probably only to him. He had been struggling for years to formulate the right ideas for understanding the course of disease and the way humans and animals become immune. Theories and experiments, ideas and facts all depend upon one another.

Because these interrelations are so complex it is easy for different thinkers to emphasize different aspects of them. Perhaps this is why there have been several rather different theories of the role of experiments in the natural sciences. I shall describe the three most important and try to show how they can be combined in an overall account.

Inductivism. The use of observation and experiment seems to mark off the scientific approach to nature from the magical or religious way of relating to the world. Impressed with this, some philosophers of science have thought that laws and theories are engendered in the minds of scientists by an intellectual process that begins with the facts experimentalists discover. It is the same facts that recommend a hypothesis to the scientific community as worthy of their belief. The process of discovery is thought to pass from the natural world of things and events, as revealed in experiments, to the ideational world of human beliefs and theories. The technical term for this supposed passage from facts to theories and laws is 'induction'. Scientists are said to arrive at their laws and theories by induction from the results of experiments, and to test them by further experiments. Observations and the results of experiments are said to be 'data', which provide a sound and solid base for the erection of the fragile edifice of scientific thought.

The inductivist theory of the role of experiments grew up slowly between the seventeenth and nineteenth centuries. Newton wrote of something like the inductivist theory in his phrase 'drawing general conclusion from experiments and observations by induction'. But he went on to say, 'and admitting of no objections against the conclusions but such as are taken from experiments, or other certain truths'. Bacon's works are probably the source of this sketch of scientific method, since it was Bacon who first saw clearly that experiments must serve the complementary functions of suggesting definitions of the nature of things and of eliminating those that are useless, by reason of their inapplicability. By the beginning of the twentieth century philosophers of science had constructed an inductivist theory of science that bore little resemblance to scientific practice, and little to the original 'induction' which Bacon had proposed. Inductivist philosophers of science thought of the laws of nature as generalizations of facts, and the accumulation of facts as supporting laws.

On reflection one can see that the inductive view must be mistaken. There are two main reasons for rejecting it, one fairly obvious, the other more subtle. First, laws and theories, in different ways, go beyond the results of experiments. Experiments are conducted here and now on just a few samples. Laws are supposed to hold everywhere and at all times and for all samples of substances. The experimental basis is too weak to support such a vast extension of scope. How can we possibly be sure that in times past or to come, and in very remote places our experiments would not have turned out very differently? And if our experiments had turned out differently so too would the laws of nature based upon them. Theories as well as laws go beyond experience. In expounding a theory scientists talk of hidden processes that produce observable effects. The pattern of iron filings that forms around a magnet can be seen, but not so the magnetic field that theory tells us is causing the filings to behave in their characteristic way. Though our knowledge of the effects of light has grown steadily and cumulatively, there have been radical changes in theoreticians' beliefs about how those effects are produced. First streams of particles were favoured, then spreading waves, and now we are back to some combination of the classical theories. How can experiments on the observable properties of man-sized material systems provide the basis for the laws of behaviour of things and processes which never could be observed by a human being?

But there is a more subtle reason why it must be wrong to think of experiments as providing the data out of which laws and theories grow. Suppose an experimentalist collects a set of data. In principle there is not just one theory which explains those data but indefinitely many from which correct descriptions of the data can be deduced. Suppose we represent the results of four experiments on a graph as in Figure 2 overleaf. Suppose that we are studying the relationship between the temperature of a gas and its volume. The crosses represent facts like 'at 20° C the volume of the gas was 30 ml.' This is the fact represented by the cross b. In the centre graph various

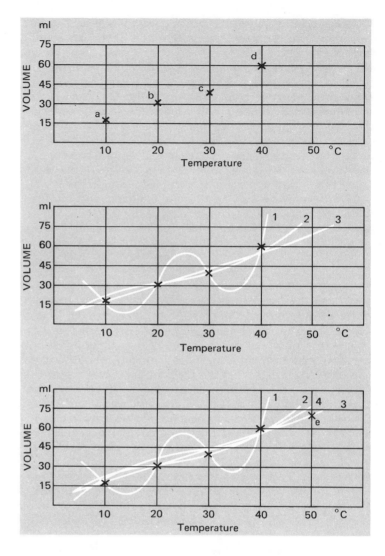

Fig. 2: *top*, the results of four
experiments; *centre*, three possible
laws; *bottom*, the effect of a fifth
experiment.

attempts to arrive at a law are represented. Each line, 1, 2 and
3, represents a possible law compatible with the data represented by the points a, b, c and d, if we allow some latitude
for error. I have shown just three possible laws, but there are
indefinitely more like them, all compatible with the data. By
doing more experiments we add more data, and so we
eliminate some possibilities. But indefinitely many more can be
added compatible with the new data. One can see this in the
diagram in the bottom graph. If we add 'e' we can eliminate
law 1. But we can easily add another law, 4, which is
compatible with all the data so far available, and there are
indefinitely many more like it.

But, it might be objected, haven't we overlooked the role of theory in science? Surely a theory could help us decide between all these laws. A simple example shows that a similar trouble infects theories. Suppose we think up a simple theory consisting of just two laws. Real theories are much more complicated but this will do to make the point. Our two laws are a theory because taken together they explain an experimentally observed finding by reference to an unobservable and more fundamental process, a process that produces, so we suppose, the phenomena we observe. The theory runs as follows:

All radio stars have strong magnetic fields.
All stars with strong magnetic fields emit X-rays.
from which we conclude

All radio stars emit X-rays.

Let us suppose that every radio star studied by astronomers to date has been observed to emit X-rays. But we could get the same conclusion from another theory.

All radio stars have high-density cores.
All stars with high-density cores emit X-rays.

From a logical point of view it doesn't matter whether the theories are true or false. Each explains the data. Unless there were to be an independent way of deciding between the rival theories, for example by finding an observable consequence of the possession of a high-density core other than the emission of X-rays, the two theories would have to stand as equally supported by the facts, at least as they were so far known. It is easy to see that there are a multitude of similar theories, all compatible with the data. This objection to a purely inductive interpretation of science is not new. It goes back to the discussions about the rival theories of the solar system that were put forward in great profusion in the sixteenth century. The problem I have just been describing was first brought into discussions of scientific method by Christopher Clavius.

Most people know of Clavius as a minor character in Brecht's *Galileo*. But he was an influential thinker in that period. In 1600 he published a textbook on astronomy, part of the intention of which was to resolve the problem of how to decide among rival theories in any branch of science, which were supported by the same data. His solution was to introduce a non-inductive criterion. He thought that theories should be judged not only for their fit to the observed or experimentally established facts, but also for their plausibility as descriptions of real but unobservable processes that cause the phenomena we observe. The issues raised by Clavius are once again central topics of discussion in philosophy of science. In subatomic physics the experimental results are very puzzling and no one theory to explain them has emerged.

Fallibilism. It has often been remarked that an experiment which fails to support a theory is sometimes more instructive than one which confirms a hypothesis. At least we know something for sure. The hypothesis from which we drew the conjecture which turned out to be mistaken must be rejected. In modern times this view of experiments has been associated with K. R. Popper. We should not think of empirical investigations as providing data which lead inexorably to laws and theories. Instead we should think of experimental results and observations as tests for laws and theories which are mere conjectures. According to the fallibilist theory of science theoreticians think of possible laws and theories, and draw out their logical consequences. These amount to predictions of what will happen in given circumstances. We know from Clavius's argument that if a prediction turns out correct the theory from which it followed might still be false. We certainly cannot say that it is true. But if the prediction fails, and assuming we know the conditions under which the law or theory from which it came are applied, that law or theory must be false. False theories, it seems to go without saying, should be rejected.

But this conception of the role of experiments suffers from its own version of the troubles that infect the inductive account. Why should a scientist reject a hypothesis that his experimental tests have shown to be false? Surely he rejects it because he expects it to be false everywhere and at all times. But how can he know that a theory that is false here today will be false in other places at other times? The world may change so that the theories which were false yesterday are true tomorrow. We cannot rule out that possibility by doing experiments. To use the results of experiments positively to prove laws rests on the unprovable assumption that the world will be similar in important ways in the future and at distant places. So too to use the results of experiments negatively to disprove hypotheses rests on the unprovable assumption that the world will *not* become *dissimilar* in important ways in the future and at distance places.

But there is a more subtle problem with fallibilism as a comprehensive philosophy of science. Laws alone do not have experimentally testable consequences. To make a prediction on the basis of a law all kinds of auxiliary hypotheses are needed, including those involved in the design of instruments. When Pasteur tested the hypothesis that the spores of anthrax bacilli were carried to the surface of the earth by earthworms, he had to assume the laws of optics because he had to trust the microscope. Failure to find the spores in the digestive tracts of worms might have been due to an unknown optical effect, just as his success in finding them depended on assuming that what he saw with the microscope was really an enlarged view of some very small things. Tests are no more conclusive when negative than when positive, since they depend on further assumptions, which might have been wrong, as to what was really responsible for an experiment failing.

Conventionalism. Both inductivism and fallibilism presume that the laws of nature are empirical statements, that is statements which are either true or false as a matter of fact. But suppose the laws of nature were neither true nor false, but were conventions for the use of words. Different sets of laws would define different ways of speaking about the world, as we come to experience it. The key question would not then be whether the laws were true or false, but under what conditions they provided the most economical, fruitful and illuminating description of reality. On this view experiments do not provide data from which laws are to be induced, nor do they serve as tests of the truth or falsity of hypotheses. The role of an experiment is illustrative. It allows a scientist to demonstrate the power of his theory, not as a collection of truths, but as a set of ideas. When an experiment succeeds, this shows that a certain way of describing the world has proved itself useful. When an experiment fails it shows that one's concepts were inadequate or confused. When one tries to describe the results of a new experiment in terms defined within an old theory, a theory which a fallibilist would say has been shown to be false, the statement by which one expresses one's attempt at description is not false but self-contradictory.

This way of looking at experiments can be illustrated from the history of chemistry. William Prout, one of the earliest biochemists, worked out a theory of atomic composition in which all atomic weights were to be integral multiples of the atomic weight of hydrogen, and so, Prout implied, all atoms were clusters of hydrogen atoms. Berzelius, taking oxygen as his standard of weight, found by experiment that the atomic weights of the elements were not integral multiples of the atomic weight of oxygen. If Prout had been right they should have been. What should we say about Berzelius's results? Had he shown Prout's hypothesis to be false? If the Proutian theory is taken as a prescription for how the term 'element' is to be used, all Berzelius had shown was that those substances which had commonly been taken to be elements were not what they seemed. Perhaps they were mixtures of more basic 'Proutian' elements. In the event the chemical world chose to accept a prescription for the use of the term 'element' in accordance with Berzelius's results, that is only those substances were to be called 'elements' which were the simplest products of chemical analysis. The issue, thus conceived, does not concern the truth or falsity of a law of nature, but the best way of prescribing the use of a term. But we should not choose to talk one way with one term, and another way with another used in the same contexts. We should have coordinated linguistic prescriptions, and these we call 'theories'.

One might imagine an analogue of this in prescriptions of terms for the offices in a social institution. The concept of 'chairman' is fixed by prescribing the duties and qualifications for the office. Calling in question the statement 'The chairman is ex-officio a member of all sub-committees' would not be to ask whether this was true as a matter of fact, but whether the office should be so defined. Different prescriptions define different institutions. One could think of a law of nature in a similar way. As prescriptions for the meaning of concepts appropriate to a possible world the laws of nature are necessary truths, conventions governing the uses of a coordinated set of concepts. Experiments could not show whether the laws were true or false. As conventions and prescriptions they do not come up for that kind of judgement. Empirical tests show whether, in this world, they are the most convenient conventions to apply.

Until modern times only one writer on scientific method managed to bring all three views together. Oddly enough it was one of the earliest thinkers to consider how scientific knowledge should best be acquired, Francis Bacon, who saw the outlines of the scientific approach most clearly. Bacon realized that the aim of experimental science is the refinement of our ideas about the natures or essences of the substances, properties and processes we find in the natural world. Typical scientific questions would be 'What is colour?', 'What is liquifaction?', 'What is heat?' In answering such questions we would have to formulate definitions of the nature of these things, processes, properties and so on. Scientific method is a disciplined and orderly way of finding answers to this kind of question.

In the preliminary stage of an investigation positive experiments and observations are assembled, correlating the effect or the substance in question with various other effects, substances and so on. Heat is found with fermentation, it is found with motion, it is found with light and with many other correlates. Each correlation suggests a hypothesis about the nature of heat. Is it a chemical effect? Is it a form of motion? Is it a radiant phenomenon? Each of these hypotheses is a possible definition of the nature of heat. In the next stage a scientist tries to falsify as many of the rival hypotheses as possible by trying to find cases of 'absence in proximity' as Bacon called it, that is cases where heat is found without fermentation, light etc. Each negative result eliminates a hypothesis. Ideally there should be only one survivor of the eliminative procedure, which would express the most powerful conception of the nature of the subject in question. In the case of heat Bacon thought it would be motion, and he defined the nature of heat as 'a motion, expansive and restrained, acting in its strife on the inner parts of bodies'.

Of course the way hypotheses are thought of and the methods by which they are tested have turned out to be very much more complicated than Bacon's somewhat primitive picture of the way to gather reliable knowledge would suggest. The elaboration of method since Bacon's time has come about because most of the natural processes, structures, properties and substances in terms of which Baconian definitions could be given have turned out not to be directly presented to the human senses. Our ideas about the hidden processes are the result of imaginative projections into the depths of nature to extremes Bacon could hardly have imagined. Nevertheless the basic logic of how we should treat a statement such as 'The proton is formed of three quarks exchanging virtual gluons' is much as Bacon sketched it when he thought about the nature of such superficial properties as heat and colour. It is a convention for the use of the word 'proton', but a convention locked into a network of concepts which recommend themselves to us in the power they have to make our experience intelligible. The world comes to seem most intelligible when the concepts with which we can understand it can be used to present a conception of the way things are in their inner natures that seems to be an accurate representation of that reality, no matter how remote from ordinary experience it may be.

What is an experiment?

A common contrast is to distinguish observations from experiments. The point of the contrast comes out in asking oneself how an observer and an experimenter stand in relation to the natural things, processes and events they study. An observer stands outside the course of events in which he is interested. He waits for nature to induce the changes, to produce the phenomena and to create the substances he is studying. He records what he has been presented with. An astronomer is the most perfect observer. He cannot manipulate the processes in the heavens. He must watch and wait. But just like an experimenter, an observer must have a well-worked out system of concepts with which to perceive, identify and describe what he sees. Without prior conceptual preparation his observations mean nothing. Perhaps the greatest scientific work based almost wholly on observation was Darwin's *The Origin of Species*. Darwin wandered round the world taking note of the plants and animals which natural processes had produced. He used the results of manipulation of nature by animal breeders and gardeners only as the basis for the analogy upon which his conception of natural selection was based. His work was a blend of theory, built up through the analogy between domestic selection and natural selection, and observations. He does not use the observations inductively, nor fallibilistically, but as illustrations of the power of his theory and its component concepts to make natural events and processes intelligible.

But an experimenter is in a different relation to natural things. He actively intervenes in the course of nature. Why should intervention be necessary? Why should nature be 'put to the question', in Bacon's phrase? In nearly all natural productions there are many processes and forces at work. Most natural effects come about through the confluence of a great many causal influences. To understand natural productions it would be advisable, if possible, to study each component causal process separately. To express these matters succinctly we need some technical terms. Experimenters describe their activities in terms of the separation and manipulation of dependent and independent variables. The independent variable is the factor in the set-up that the experimenter manipulates directly. The dependent variable is the attribute which is affected by changes in the independent variable. A cook can control the amount of chilli in the curry (an independent variable), and thereby affect the amount of water consumed by the diners (the dependent variable). But in the real world there are hardly any processes so simple that they can be manipulated by one variable representing a cause and another its invariable effect.

By careful design of an experiment it is possible to maintain constant all properties except those one wishes to study, the dependent and independent variables. A property which is fixed in this way is called a 'parameter'. Fixing the parameters defines the state of the system within which the variables are to act. Many of the experiments in this book depended on the skill of the experimenters in fixing parameters. For instance in their experiments to measure the 'spring of the air' Boyle and Hooke kept the temperature of the trapped air constant. Later experimenters, such as Amagat and Andrews, repeated Boyle's experiment at different fixed temperatures. They found that different laws obtained with different values of the parameter. Sometimes Boyle used pressure as an independent variable, sometimes he manipulated volume and measured the consequent change in pressure.

The need to separate the variables and to fix parameters seriously restricts the use to which experiments can be put. There are many phenomena, particularly in the world of human action, in which the practical separation of variables and parameters cannot be managed. This is because attempts at isolation simply change or even destroy the property one wishes to study. For instance in social studies one must allow for the context within which a human action occurs, since how an action is interpreted is determined by its context, and the context in turn determines the effect it is likely to have. A smile, for example, can mean many different things depending on all the other actions which precede and accompany it. A certain smile may suggest anything from reassurance to threat depending on its context and accompaniments. So there could never be experiments on the effect of smiling, in which the smile was taken as an independent variable, its effects on others as the dependent variable and the situations in which smiling occurred fixed as parameters.

However, there is another kind of intervention in the natural world which yields knowledge, but lacks the manipulative character of the true experiment. I shall call this kind of intervention an 'exploration'. An anatomist is not experimenting when he dissects an animal or plant, nor is a geologist when he charts the structure of the earth's crust. There are intermediate procedures, part experiment, part exploration. For instance the use of X-ray diffraction to study the structure of crystals requires manipulations very similar to those used in true experiments. I have not included any pure explorations in this collection, though Aristotle's experiment with the clutch of eggs has a strongly exploratory character.

What sort of matters can be studied experimentally?

By far the commonest sort of experiment must surely be the measurement of some variable property under differing conditions. One studies the change in the electrical conductivity of molten potassium chloride with changes in temperature. The results is a mathematical function linking the two variables, say k and θ. It might look like this:

$$k = a\theta^2$$

where k is the electrical conductivity, θ the temperature and a a constant. The experiments that led to Boyle's Law exemplify this kind of study to perfection. Manipulations like this can easily be extended to identify the limits within which a law holds. Does Boyle's Law continue to hold good at very high pressures or at very low temperatures, or with gases much denser than air? This kind of question can be pursued simply by widening the range of the variables with which one is experimenting, going to higher and higher temperatures, for instance.

Perhaps the next commonest kind of experiment involves the attempt to relate the structure of things, discovered in an exploratory study, to the organization this imposes on the processes going on in that structure. Hales's efforts to find out about the circulation of fluids in plants were based upon Nehemiah Grew's explorations of the structure of the stem, and his discovery that it is made up of continuous, fluid-filled vessels.

Less common than either of these types, but often the most helpful in testing a theory, are experiments which reveal the existence of something not previously identified in the real world. Sometimes expectations of the existence of something are not fulfilled. Usually there is a deliberate search inspired by a theory. A prior specification of what the thing, substance or process is likely to be like guides the research. There are several examples of experiments of this kind in this book. Davy's successful separation of the alkali metals depended not only on the bold extension of a technique, but on his having a pretty good idea what he was likely to find by using it.

Fig. 3: *A Chemist's Laboratory*, signed and dated 'L R 1827'. Oil painting in the Museum of the History of Science, Oxford University. The picture is said to represent Sir Humphry Davy with an assistant.

Instruments

In the romantic view of the experiment the apparatus and equipment loom large. Glittering glassware and mysterious meters are the focus of aesthetic interest. But why do scientists need equipment to study the natural world? One can begin to answer this question by distinguishing three kinds of instruments. There is equipment for making measurements: clocks, meters, graduated rules and so on. Then there is the apparatus for extending the human senses: microscopes, telescopes, amplifiers, stethoscopes etc. But at the heart of the experiment is the equipment that enables an experimenter to isolate the effect he wishes to study, and to separate the possible causes of it.

Instruments for making measurements and pieces of equipment that extend the human senses depend on certain assumptions and beliefs about their relations to the things in the world. Consider a simple graduated rule that might be used to measure the length of a metal rod. In taking the result of the measurement as the length of the rod, one has to make a number of physical assumptions. The end of the rod and the relevant mark on the rule have to be judged to be coincident. But to accept the eye's verdict one has to assume that rays of light travel in straight lines from their source to the eye. More recondite assumptions are involved when the measuring operations require the rule to be moved. The rule must not shrink or expand as it is shifted along the side of the thing to be measured. All this may seem terribly obvious, and scarcely worth the expenditure of ink to point it out. But when the measuring equipment is in motion relative to the thing measured it has turned out that many of our common-sense assumptions are just plain mistaken. If a measuring device is moving past a stationary (relatively stationary) thing it 'shrinks' in the direction in which it is moving, so that stationary object will seem to be longer than if it had been measured by equipment which was also stationary. Quite subtle physics is required to make allowance for these phenomena, and further thought on these matters has led physicists to query the assumption that we can properly talk of *the* length of something. I shall illustrate some of the issues involved in describing the attempt by Michelson and Morley to measure the speed of the earth's passage through space.

Microscopes and telescopes are typical instruments for extending our ordinary senses. But if we are to believe that they reveal good views of hidden denizens of the real world, whether they are small like bacteria or very large like distant stars, we must assume a great deal of the physics of light. Again, if one is using a simple magnifying glass to examine the water in a pond this point hardly seems of great moment, after all one can almost see the paramecia with the naked eye. But when one is examining the very small and the very distant the physics plays a larger part. For instance, in using a telescope to examine galaxies that are very far away astronomers noticed that the light was much redder than they would have expected, and indeed redder than light they received from similar objects which they took to be much closer. Physics tells us that if something is moving away from us, the faster it recedes the

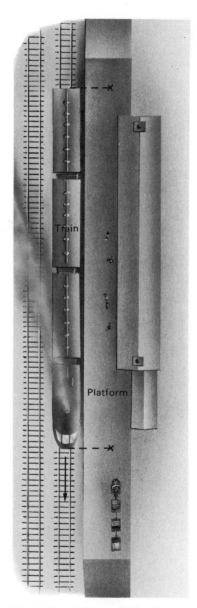

Fig. 4: Einstein illustrated the effect on measurement of apparent shrinkage in the direction of motion with an observer on a stationary platform measuring the length of a moving train.

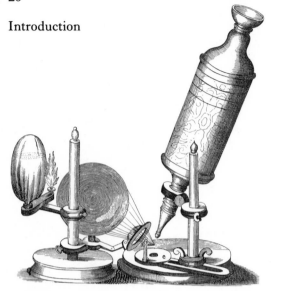

Figs 5 and 6: The two basic
instruments for extending our senses.
Left, a microscope (*c*.1664) used by
Robert Boyle's assistant and
collaborator, Robert Hooke. *Right*, a
nineteenth-century reflecting
telescope.

redder its light will be. Astronomers were presented with a
problem of interpretation. Were the reddish objects they saw
the same distance away as similar but bluer stars, only emitting
redder light? Or were they emitting light of the same wavelength as nearer things but moving away at speed? For a
variety of reasons cosmologists chose the latter solution. We
now speak confidently of the 'red shift' and the 'expansion of
the universe'. But our instruments do not reveal these phenomena. They are brought into being by an act of interpretation
based on physical theory.

Whether one is dealing with a very simple instrument
involving little by way of interpretation or with equipment
related in more complex ways to the phenomena we take it to
reveal, our willingness to accept the deliverances of the
instrument as a proper record of some natural event depends
on our faith in the causal relations that obtain between the state
of affairs in the world and the effect it has on the instrument.
The thermometer is a good example. A simple causal relation
links the degree of heat of the material being measured with the
expansion that that degree of heat induces in the liquid
enclosed in the tube. The greater the degree of heat the greater
the expansion of the liquid, and so the longer the column of
liquid in the stem of the thermometer. In most instruments in
common use there is a fairly simple relation between the state
of the instrument and the state of the world it is used to
measure. Even if the physical relation is complex most
instruments take up a definite state when acted upon by the
world which is also in a definite state. One might call such

instruments 'transparent'. It used to be thought that with a little ingenuity all instruments could be made transparent. If an instrument is thrown into a definite state by a specific state of the system it is measuring, and if the same state of the world always produces the same state in the instrument, it will always be possible to infer the state of the world from the state of the instrument. This is called the 'faithful measurement postulate'.

Unfortunately the faithful measurement postulate does not always hold. When sociologists tried to emulate what they took to be the methods of the physical sciences they introduced questionnaires as the analogues of instruments. They thought they could 'measure' people's attitudes and beliefs. But they overlooked the faithful measurement postulate. Questionnaires are not transparent, since it would be unwise to assume that one can infer a respondent's attitudes from his answers to the questions put to him in an interview. People want to appear in the best possible light to an investigator, even if it is one they never meet face to face. People say things for a bewildering variety of motives, which differ from person to person, and from one moment to another. Similar troubles have beset instrumentation even in physics. If one does a series of experiments preparing subatomic particles in exactly the same way each time, the results will generally be different. No matter how determinate the preparation of the beam of particles, there is a scatter of different results. If a beam of electrons is sent through a small hole they do not all strike the same spot on a screen. When a photographic plate is used as a detector a characteristic scatter pattern can be observed. In the Stern–Gerlach experiment we have a simple case of the phenomenon. We can say in advance what are the possible states a particle can take up in a certain apparatus, that it will follow either a left-handed or a right-handed path. However, we cannot say which path any particular particle will follow. All we know is that in the long run half the particles will take up one state and half the other.

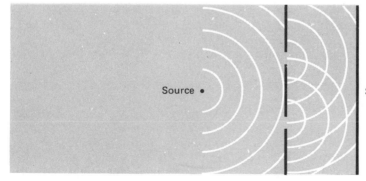

Source •

Screen

Fig. 7: Interference pattern of waves: beams of electrons, projected through slits in a solid screen, behave like waves, precluding the exact measurement of all the physical properties of single electrons.

But more important than measurement and the extension of the senses is the role of equipment in isolating influences and tendencies, allowing each to be studied independently. How is this possible? Setting up an experimental apparatus is essentially a way of creating an isolated environment. In the simplified world created in the apparatus the properties that one wishes to study can be manipulated. On page 22 I introduced the terminology of independent and dependent variables, to describe this kind of experiment. It is hoped that the apparatus is so arranged that all outside influences are either eliminated or controlled, that is kept constant as parameters. By floating their equipment in a bath of mercury Michelson and Morley were able to isolate their apparatus from the vibrations and other disturbances that emanated from the city of Cleveland. Sometimes, instead of trying to eliminate external influences, they can be controlled so that they always bear on the apparatus in the same way. In increasing pressure on their enclosed air Boyle and Hooke caused it to warm up slightly, so they allowed the compressed gas to cool again to room temperature. They could not eliminate the effect of temperature but by maintaining it constant they could assume that its effect would be always the same. Sometimes the elimination of a factor is built into the phenomenon, so to speak. Theodoric did not need to ensure that his water-filled flasks which simulated raindrops fell with a constant acceleration, like real raindrops. He realized that drops replaced each other in the curtain of rain so quickly that for all practical purposes they were stationary.

The twenty case histories which now follow are chosen to illustrate the points I have been making about the nature of experiments and the roles they play in the acquisition of scientific knowledge. But I have not lost sight of the romantic aspect of experimental science. I hope that these accounts will be read as illustrations of human skill and ingenuity, and that each experiment will be seen, each in its own way, to be something of a work of art.

I

Formal Aspects of Method

A

Exploring
the Characteristics of a
Naturally Occurring Process

The simplest way in which a scientist can actively seek knowledge is deliberately to exploit a natural process, but a process which he cannot control. In this section I describe two investigations, the one by **Aristotle** on the embryology of the chick, and the other by **William Beaumont** on the process of digestion. In both cases a natural process was isolated and systematically observed; but its unfolding was not able to be controlled.

1. Aristotle

The Embryology of the Chick

Aristotle was born in Stagira, a Greek colony in Asia minor, in 384 BC. His father was a doctor, a member of the guild of the Asclepiadae. Aristotle was orphaned while still a child, and brought up by a relative. It does seem likely that even while very young he had some training in medical and biological matters from his father.

At the age of eighteen he entered Plato's Academy at Athens, and remained there until Plato's death in 347 BC. As a young man he seems to have cut something of a figure. Anecdotes about this period in his life suggest that he attracted a certain amount of envy for his stylish manners and intellectual advantages, a combination of qualities hard to forgive in any age. After Plato's death he left Athens for Atarneus. This was a small state whose ruler, Hermias, had collected a circle of scholars influenced by Plato's teachings. Shortly after his arrival Aristotle married Hermias's adopted daughter, Pythias. They had only one child, a daughter called after her mother. After his wife's death Aristotle set up house with a woman called Herpyllis, though it seems he never married her. Nicomachus, their son, was the recipient of the moral treatise from his father that has come down to us as the *Nicomachean Ethics*.

Aristotle stayed at Atarneus for three years, and then moved to Mytilene on the island of Lesbos. It seems likely that he made most of his biological investigations while living there. Sometime in 343–342 he was invited to tutor Alexander, the son of Philip of Macedon. Eight years later he returned to Athens and founded his own school and library, the Lyceum. Schools like the Academy and the Lyceum served some of the functions of modern universities, though they were not so formally organized.

By 322 feeling had turned against the Macedonians and Aristotle retired to Chalcis. He remarked that he did not want

Fig. 8: Portrait bust of Aristotle. Rome, Terme Museum.

to give the Athenians a chance to destroy another philosopher, as they had Socrates. He died in Chalcis shortly afterwards.

Theories of organic generation before Aristotle

With Darwin, Aristotle must surely be ranked as among the greatest biologists. He was one of the very first to carry out systematic observations and to write a detailed work on organic forms, known to us as the *Historia Animalium*. The experiment I shall be describing laid the foundations for all subsequent embryological work. It is remarkable both for its systematic character, and for the shrewdness of the questions Aristotle was prompted to ask by the results of his investigations.

The problem of the nature of 'generation', the way animals and plants came into existence, had been quite deeply considered by Greek thinkers before Aristotle. How does a new plant or animal come into being? It seems to be formed out of some basic undifferentiated stuff, and yet it quickly takes on a most refined and articulated structure. Is that structure just a filling out of a pre-existing plan (the theory of pre-formation), or does it come into being stage by stage, as the various phases of the growth process unfold (the theory of epigenesis)? The problem is not wholly solved even today. Attempts to understand the process of generation are very ancient, and already in 345 BC Aristotle was the inheritor of a body of doctrine from a long line of predecessors interested in the problem.

The only medical treatises of worth to come down to us from the times before Aristotle are the Hippocratic writings. Whoever wrote these works had a very clear idea of the possibilities of comparative embryology of non-human species as an approach to the problem of how new human beings are created. In the work *On the Nature of the Infant* an exploratory study is suggested in the clearest terms. 'Take twenty eggs or more, and set them for brooding under two or more hens. Then on each day of incubation from the second to the last, that of hatching, remove one egg and open it for examination. You will find that everything agrees with what I have said, to the extent that the nature of a bird ought to be compared with that of a man.' Commentators on these writings seem to be agreed that the text does not suggest that the author actually followed his own prescription. That was left to Aristotle. Here is his description of the embryonic stages in the development of the chick.

The opening of the eggs

'Generation from the egg proceeds in an identical manner with all birds, but the full periods from conception to birth differ, as has been said. With the common hen after three days and three nights there is the first indication of the embryo; with larger birds the interval being longer, with smaller birds shorter. Meanwhile the yolk comes into being, rising towards the sharp end, where the primal element of the egg is situated, and where the egg gets hatched; and the heart appears, like a speck of blood, in the white of the egg. This point beats and moves as though endowed with life, and from it two vein-ducts with blood in them trend in a convoluted course [as the egg-substance goes on growing, towards each of the two circumjacent integuments]; and a membrane carrying bloody fibres now envelops the yolk, leading off from the vein-ducts. A little afterwards the body is differentiated, at first very small and white. The head is clearly distinguished, and in it the eyes, swollen out to a great extent. This condition of the eyes lasts on for a good while, as it is only by degrees that they diminish in size and collapse. At the outset the under portion of the body appears insignificant in comparison with the upper portion. Of the two ducts that lead from the heart, the one proceeds towards the circumjacent integument, and the other, like a navel-string, towards the yolk. The life-element of the chick is in the white of the egg, and the nutriment comes through the navel-string out of the yolk.

When the egg is now ten days old the chick and all its parts are distinctly visible. The head is still larger than the rest of its body, and the eyes larger than the head, but still devoid of vision. The eyes, if removed about this time, are found to be larger than beans, and black; if the cuticle be peeled off them there is a white and cold liquid inside, quite glittering in the sunlight, but there is no hard substance whatsoever. Such is the condition of the head and eyes. At this time also the larger internal organs are visible, as also the stomach and the arrangement of the viscera; and the veins that seem to proceed from the heart are now close to the navel. From the navel there stretch a pair of veins; one towards the membrane that envelops the yoke (and, by the way, the yolk is now liquid, or more so than is normal), and the other towards that membrane which envelops collectively the membrane wherein the chick lies, the membrane of the yolk, and the intervening liquid. [For, as the chick grows, little by little one part of the yolk goes upward, and another part downward, and the white liquid is between them; and the white of the egg is underneath the lower part of the yolk, as it was at the outset.] On the tenth day the white is at the extreme outer surface, reduced in amount, glutinous, firm in substance, and sallow in colour.

The disposition of the several constituent parts is as follows. First and outermost comes the membrane of the egg, not that of the shell, but underneath it. Inside this membrane is a white liquid; then comes the chick, and a membrane round about it, separating it off so as to keep the chick free from the liquid; next after the chick comes the yolk, into which one of the two veins was described as leading, the other one leading into the enveloping white substance. [A membrane with a liquid resembling serum envelops the entire structure. Then comes another membrane right round the embryo, as has been described, separating it off against the liquid. Underneath this comes the yolk, enveloped in another membrane (into which yolk proceeds the navel-string that leads from the heart and the big vein), so as to keep the embryo free of both liquids.]

About the twentieth day, if you open the egg and touch the chick, it moves inside and chirps; and it is already coming to be covered with down, when, after the twentieth day is past, the chick begins to break the shell. The head is situated over the right leg close to the flank, and the wing is placed over the head; and about this time is plain to be seen the membrane resembling an after-birth that comes next after the outermost membrane of the shell, into which membrane the one of the navel-strings was described as leading (and, by the way, the chick in its entirety is now within it), and so also is the other membrane resembling an after-birth, namely that surrounding the yolk, into which the second navel-string was described as leading; and both of them were described as being connected with the heart and the big vein. At this conjuncture the navel-string that leads to the outer after-birth collapses and becomes detached from the chick, and the membrane that leads into the yolk is fastened on to the thin gut of the creature, and by this time a considerable amount of the yolk is inside the chick and a yellow sediment is in its stomach. About this time it discharges residuum in the direction of the outer after-birth, and has residuum inside its stomach; and the outer residuum is white [and there comes a white substance inside]. By and by the yolk, diminishing gradually in size, at length becomes entirely used up and comprehended within the chick (so that, ten days after hatching, if you cut open the chick, a small remnant of the yolk is still left in connexion with the gut), but it is detached from the navel, and there is nothing in the interval between, but it has been used up entirely. During the period above referred to the chick sleeps, wakes up, makes a move and looks up and chirps; and the heart and the navel together palpitate as though the creature were respiring. So much as to generation from the egg in the case of birds.'

(*Historia Animalium*, book 6, 561a3–562a20)

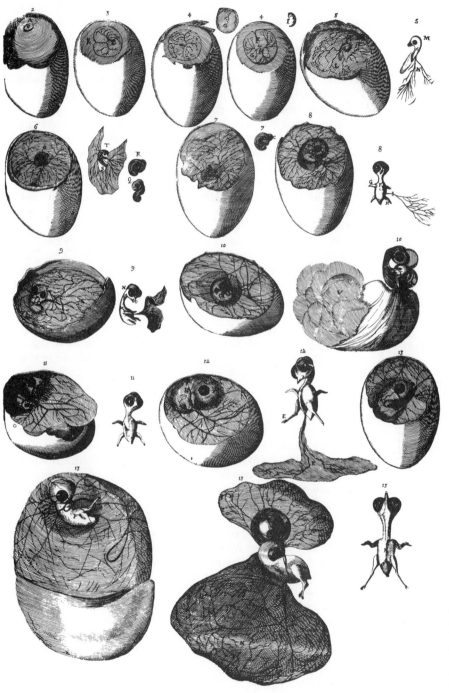

Fig. 9: Embryo
chicks at different
stages of de-
velopment. En-
graving from H.
Fabricius, *De
Formatione Ovi et
Pulli*, Padua (1621).

Aristotle

No doubt interest in embryology continued after Aristotle's time, particularly in widening the scope of observational and experimental studies. But very little of the work of Hellenistic science, from the great schools of Alexandria, has come down to us. Medieval Europe learned most of its Greek science from Arabic authors, who had transmitted and enlarged the ancient learning. Amongst the most important sources of medical and biological knowledge were the works of Galen and Avicenna. But medieval science, for the most part, returned to Aristotle as an ultimate source, so that new work was usually the result of critical commentaries on surviving Aristotelian treatises. In particular medieval embryology was closely modelled on the section I have quoted from Aristotle's *Historia Animalium*.

One of the most sophisticated treatises on generation, in the Aristotelian tradition, was composed by Giles of Rome about 1276. In this work, *De Formatione Corporis Humani in Utero*, there are theoretical discussions of the relative contribution of the male and female parent to the generative process. There are detailed descriptions of foetal development extending Aristotle's study of the development of embryo birds to include human development. Giles's treatise attracted a good deal of criticism, very revealing about the growth of embryological knowledge in the Middle Ages. According to Hewson, criticisms by James of Forli and Thomas del Garbo of Giles's description of the membrane surrounding the embryo point to the use of authorities other than Aristotle, particularly in works of Arabic origin.

The issue centred on the disposition, function and order of development of the three embryonic membranes. It seems clear that the criticism of Giles's descriptions owes something to dissection as well as to the use of new authorities. The order of development of the membranes may seem to be a matter of little importance, but it was connected with the controversy between pre-formationists and epigeneticists, a controversy that goes back to the earliest Greek sources.

In drawing on Galen's writings, Giles had to hand a much more detailed source than anything to be found in the works of Aristotle. But there was no scientific revolution in the history of embryology. Successive observers improved the quality and accuracy of their descriptions, refining and correcting the traditional wisdom. In his *De Formato Foetu* of 1604 Fabricius describes very much the same structures as Aristotle had recorded, and discusses very much the same problems as had bothered Giles of Rome. All agree that the foetal membranes serve the dual function of protecting the embryo and storing waste. Each realized that the pace of foetal development is best studied by referring all other sequences to the development of

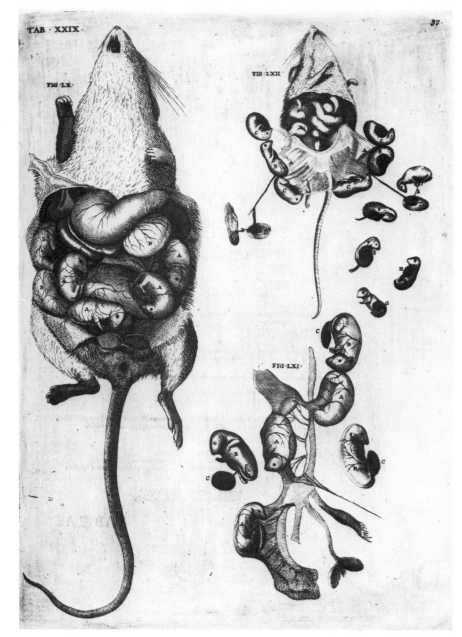

Fig. 10: Embryo rats
showing sixteenth-
century
developments in
understanding the
blood supply to the
foetus. Engraving
from H. Fabricius,
De Formato Foetu,
Padua (1604), table
xxii.

the blood vessels. Fabricius added a detailed description of the
blood system of the umbilical cord, contributing one more
brick to the growing edifice of knowledge.

In reading Aristotle's description one must surely be struck
both by the clarity of the account, reflecting the care with

which the various stages were observed, and by his obvious grasp of the main physiological principles involved, particularly the distinctive roles of the white and the yolk. Already in the comparison between the membranes and the mammalian after-birth Aristotle is generalizing his embryological observations from one species to others.

But in what sense is this study an experiment? I distinguished empirical investigations which explore the given things and processes of nature from those in which active intervention is used to isolate causal influences and identify their particular effects. Greek science was largely exploratory and theoretical. But in the controlled use of the sequence of eggs we have an example of an investigative technique which involves some interference and some contrivance. Aristotle did not wait passively for the stages of development of the chick to be presented to him, but actively intervened in the natural process in the ingenious way suggested by the Hippocratic author.

Further reading

Aristotle, *Historia Animalium*, transl. D. W. Thompson, Oxford, 1910.

Adelman, H. A., *The Embryological Treatises of Hieronymus Fabricius*, Ithaca, N.Y., 1942, vol. I, p. 37.

Allan, D. J., *The Philosophy of Aristotle*, 2nd edn., Oxford, 1970.

Hewson, M. A., *Giles of Rome and the Medieval Theory of Conception*, London, 1975.

2. William Beaumont

The Process of Digestion as Chemistry

William Beaumont was the son of a farmer, born in Lebanon, Connecticut, in 1785. Being of a somewhat adventurous disposition he left home in 1806, 'with a horse and cutter, a barrell of cider and $100'. His first settled employment was as a schoolmaster in Champlain, New York, in 1807. During his stint in the schoolhouse he borrowed books on medicine and read widely in the associated sciences. He apprenticed himself to Dr B. Chandler of St Albans, Vermont, in 1810, and two years later received his licence to practice. He joined the U.S. Army in 1812 during the war with Britain, and stayed on till 1815. He practised in Plattsburg, Pennsylvania, until 1820, when he rejoined the U.S. Army with a commission. He was posted to Fort Mackinac in the Michigan area.

It was there that the accidental injury to an Army servant occurred upon which Beaumont's great experimental programme was dependent, and which will be described in this section.

Beaumont seems to have been tolerably happy in the Army, and he stayed on in various posts until 1839. His studies on the chemistry of digestion had become internationally famous in those years, particularly in Germany, where he was influential on such workers as Johannes Müller.

His last posting was to St Louis, and it was there that, on leaving the Army, he set up in practice. In 1853 he suffered a severe fall from a horse. He died shortly afterwards from the subsequent infection.

Early work on digestion

The most sophisticated studies of digestion prior to those of the nineteenth century were the work of J. B. van Helmont, a Flemish doctor. He was a man of great originality of thought, and with the manipulative skill and ingenuity to carry out

Fig. 11: William Beaumont, portrait in the National Library of Medicine, Bethesda, Maryland.

William Beaumont

empirical studies, and even experiments to test (or rather to demonstrate) his theories of digestion. Most of his work is summed up in a strange but immensely popular work, the *Oriatrike or Physic Refined*, published in English translation in 1662. Like all good scientists, he clears the ground of palpably mistaken theories prior to recommending his own. In van Helmont's day, most people thought of digestion as a kind of cooking brought about by the heat of the stomach. A simple observation was enough for him to dispose of the 'coction' theory – 'for therefore, in a fish, there is no actual heat, neither therefore notwithstanding, doth he digest more unprosperously than hot animals.' Cold-blooded fish digest their food as well as hot-blooded animals.

Van Helmont is credited with the first alkaline prescription for the cure of indigestion, a treatment he based upon his observations of the acidity of the stomach juices. 'I oftentimes', he says, 'thrust out my tongue, which ... [a tame] Sparrow laid hold of by biting and endeavouring to swallow to himself, and then I perceived a great sharpness to be in the throat of the Sparrow, whence from that time I knew why they are so devouring and digesting.' But acid is not sufficient for digestion. He proved this by showing that vinegar will not dissolve meats. There must also, he argued, be 'Ferments', which are specific in their actions for different classes of foods, 'for mice ... do sooner perish of hunger than eat of a ringdove'. Van Helmont's notion of a Ferment is very near to our modern concept of an enzyme. Not only did he believe that there were Ferments in the stomach and duodenum (which

Fig. 12: Fort Crawford, Prairie du Chien, Wisconsin. This is one of several forts where Beaumont was posted while he continued his experiments. Photo National Library of Medicine, Bethesda, Maryland.

latter organ he knew to contain alkaline juices), but also each
organ had its own specific enzymes or Ferments 'where the
inbred-spirit in every place doth cook its own nourishment for
itself'.

Little further advance had been made in the experimental
study of digestion in the years intervening between the studies
reported by van Helmont and those undertaken by William
Beaumont. This reflects the advanced character of van Helmont's concepts, rather than any backwardness in biochemical
studies. Not only had van Helmont introduced essentially the
modern concept of an enzyme, but it was he who first proposed
an 'invasion' theory of disease, ancestor of the bacterial theory.
He held that illness was caused by the invasion of the body by
alien *archeae*, which took over the life processes for their own
advantage, releasing poisonous waste products which are the
immediate causes of the symptons of common illnesses. Van
Helmont was immensely revered among medical men.

The St Martin Experiment

On 6 June 1822 a certain Alexis St Martin, a French Canadian
serving as a porter and general servant with the Army, was
wounded in the abdomen by a musket, accidentally discharged
at very close range. St Martin was only eighteen but of a most
robust constitution. When he was brought to Beaumont the
surgeon found that there were several serious wounds including perforation of the abdominal wall and the stomach.
Through this hole 'was pouring out the food he had taken for
breakfast'. St Martin must have had a remarkable physique,
since when he developed a fever from infection in the wound
he was 'bled to the amount of 18 or 20 ounces . . .' According to
Beaumont, 'the bleeding reduced the arterial action and gave
relief'(!)

Gradually the wound healed. At first St Martin could keep
no food in the stomach, but 'firm dressings were applied and
the contents of the stomach retained.' Beaumont reports that
'after trying all the means in my power for eight or ten months
to close the orifice . . . without the least success . . . I gave it up
as impractical.' Within eighteen months a small fold or
doubling of the coats of the stomach 'appeared forming at the
superior margin of the orifice, slightly protruding and increasing till it filled the aperture, so as to supersede the necessity for
the compress and bandage for retaining the contents of the
stomach.' This 'valve' was easily depressed with the finger. At
about this time it seems to have suddenly dawned on Beaumont that in St Martin and his peculiar injury there was an
ideal laboratory for an experimental study of digestion. The
French Canadian was an exceedingly tough man. Beaumont
reports that during the whole time that he used St Martin in

William Beaumont

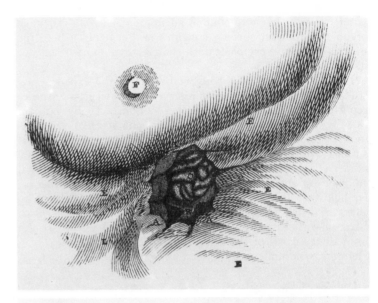

Fig. 13: The wound at an early stage. Illustration from Beaumont, *Experiments and Observations on the Gastric Juice and the Physiology of Digestion*, Edinburgh (1838), p. 17.

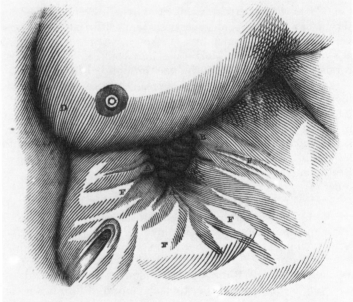

Fig. 14: Folding in to form a natural valve. Beaumont, *Observations* (1838), p. 19.

these studies he was generally in good health and active, athletic and vigorous. Their curious partnership persisted for nine years. There were only occasional interruptions as St Martin returned to Canada, married, and from time to time took up other occupations. In 1833 Beaumont remarks that 'for the last four months he [St Martin] has been unusually plethoric and robust, though constantly subjected to a continuous series of experiments on the interior of the stomach.'

The work divided into two interlocked series of experiments. In one series various substances were studied as they were digested in the natural conditions of the stomach, an experiment *in vivo*. In the other, stomach juices were extracted and the conditions for their action on food materials studied outside the body, an experiment *in vitro* (in a glass vessel). The whole of the work Beaumont carried out over the period of his association with St Martin could be thought of as one great experiment, systematically varying the conditions under which digestion occurred to discover what was really crucial to its proper functioning. But it could also be looked upon as a series of independent 'experimentules', small-scale events each of which contributed to the overall understanding of the process.

It was easy enough to drain out the digestive ferments, 'by placing the subject on his left side, depressing the valve within the aperture, introducing a gum elastic tube and then turning him . . . on introducing the tube the fluid soon began to run.' The chemistry of the duodenum could be studied *in vitro* too because 'bright yellow bile can also be obtained flowing freely through the pylorus . . . by pressing the hand upon the haptic region.' And when some food has been digested in the stomach 'the chymous fluid can easily be taken out . . . by laying the hand over the lower part of the stomach . . . and pressing upwards.'

The basic studies concerned the rate and temperature of digestion, and the chemical conditions that favoured it at different stages of the process. In the course of the experiments Beaumont noticed the marked way the stomach lining was injured and became morbid through any kind of indulgence or improper feeding. He remarks that 'improper indulgence . . . eating and drinking, has been the most common precursor of these diseased conditions of the coats of the stomach . . . but seldom indicated by any ordinary symptom or particular sensation.' That St Martin was somewhat self-indulgent from time to time can be read off from the second of the tables reproduced here. Beaumont summed up his results in tables, in which the digestive process in the stomach is compared with that which can be artificially induced by the use of gastric juices in glass containers maintained at suitable temperatures.

A typical experiment examining digestion outside the body went as follows: 'February 7. At 8 o'clock, 30 minutes, A.M. I put twenty grains *boiled codfish* into three drachms of gastric juice and placed them on the bath.

At 1 o'clock, 30 minutes, P.M., fish in the gastric juice on the bath was almost dissolved, four grains only remaining: fluid opaque, white, nearly the colour of milk. 2 o'clock, the fish in the vial all completely dissolved.'

Corresponding experiments were carried out *in vivo*. Again an experiment typical of hundreds went as follows: 'At 9 o'clock he breakfasted on *bread*, *sausage* and *coffee*, and kept exercising. 11 o'clock, 30 minutes, stomach two-thirds empty, aspects of weather similar, thermometer 29° [F], temperature of stomach 101½° and 100¾°. The appearance of contraction and dilation and alternate piston motions were distinctly observed at this examination. 12 o'clock, 20 minutes, stomach empty.'

TABLE,   269

Showing the mean time of digestion of the different Articles of Diet, naturally, in the Stomach, and artificially, in Vials, on a bath.

The proportion of gastric juice to aliment, in artificial digestion, was *generally* calculated at one ounce of the former to one drachm of the latter, the bath being kept as near as practicable at the natural temperature, 100° Fahrenheit, with frequent agitation.

Articles of Diet.	Mean time of chymification			
	In Stomach.		In Vials.	
	prep.	h. m.	prep.	h. m.
Rice, -	boiled	1 00		
Sago, -	do.	1 45	boiled	3 15
Tapioca, -	do.	2 00	do.	3 20
Barley, -	do.	2 00		
Milk, -	do.	2 00	do.	4 15
Do. -	raw	2 15	raw	4 45
Gelatine. -	boiled	2 30	boiled	4 45
Pig's feet, soused,	do.	1 00		
Tripe, do.	do.	1 00		
Brains, animal,	do.	1 45	do.	4 30
Venison, steak,	broiled	1 35		
Spinal marrow, animal,	boiled	2 40	do.	5 25
Turkey, domesticated,	roasted	2 30		
Do. do.	boiled	2 25		
Do. wild,	roasted	2 18		
Goose, do.	do.	2 30		
Pig, sucking -	do.	2 30		
Liver, beef's, fresh,	broiled	2 00	cut fine	6 30
Lamb, fresh,	do.	2 30		
Chicken, full grown,	fricas'd	2 45		
Eggs, fresh,	h'rd bld	3 30	h'rd bld	8 00
Do. do.	soft bld	3 00	soft bld	6 30
Do. do.	fried	3 30		
Do. do.	roasted	2 15		
Do. do.	raw	2 00	raw	4 15
Do. whipped.	do.	1 30	whipped	4 00
Custard, -	baked	2 45	baked	6 30
Codfish, cured dry,	boiled	2 00	boiled	5 00

Fig. 15: 'Table, showing the mean time of digestion of the different Articles of Diet, naturally, in the Stomach, and artificially, in Vials, on a bath'. From the original edition of Beaumont's *Observations*, Plattsburg (1833), p. 269.

Though these experiments taken together provide a marvellous descriptive account of the times and conditions for the digestion of a wide variety of common foods, they were also seen by Beaumont and his contemporaries as bearing most directly on a theoretical controversy of some antiquity and importance. The problem can be summed up in an apparently simple question: 'Is the gastric juice a *chemical* solvent?' The alternative theory required that there be some special vital force present in living organisms and needed in the digestive

Date 1829	Wind and Weather	Fh.	Empty: repose	Empty: ex'rcis	Dur'g chym'n: repose	Dur'g chym'n: ex'rcis	
Dec 6	N	Cl'dy and damp	63	98°				
7	N W	do do	27	98				
8	S W	Clear and dry	13	99				
9	W	Clear	10	99				
1830 Jan 21	N W	do and cold	0.8	100				
25	S W	do	2	100		100		
Mar 17	S W	Rainy	33	98			102	
18	N W	Clear	6	100				
9				98				
1832 Dec 4	N W	Snowing	35	100	101		101 1 2	
5			30	100				
6			38	100				
7			28	99		100		Stomach morbid. do do
8		Cl'dy and damp	46	99		99		do do
13				100				
14				100				Stomach morbid. do do
15				100				
22				100		100		
23				100	101			Stomach morbid. do do
25	E	Variable	31	100		100	101	
26	N E	Cl'dy and damp	38	99 1-2	101	99 1 2	101	
27	E	Foul and damp	33	99 1 2		100		
"	S	Clear	62	100		100		
28	N	do	31	100				
29	N W	do	34	100		100		
30	"	do	26	100				
31	S	Cl'dy and damp	30	100 1 2				Stomach morbid.
1833 Jan 1	S	Rainy	50	100				
3		Clear	38		101 1.2			
7	N E	Cl'dy and damp	48	100				
11	S W	Clear	15	100				
13	C'lm	Cloudy and dry	12	100	101	100	100 1 2	Stomach morbid.
14	N W	Clear	23	100			101 1 2	
15	N E	Cloudy and dry	35	100	101			
17	N W	Clear and dry	19	100		100	102	stomach morbid.
23	N E	Rainy	39	100 1 2			101 3.4	
24	N	Cl'dy and damp	39	100 1 2	101 1'4			after sleeping.
"	N E	Rainy		99 1 2				before rising.
25	S		36	99			102	
"			33	100 1.2				
26	N W	Clear	36	100 1 2		100 3.4	101	99 1 2 aft. sl'p'g.
27	C'lm	Cloudy	32	99 1 2			101 1.4	99 1 2 bef. rising.
28	S W	Clear	35	101 *			101 1.2	
"	S W	do	46	101 1 2		101 1 2		
..				101 1.2		101 1 2		
29	N E	Clear	28	100 3 4	101 1.2		102	100 before rising.
30	N E	Cl'dy and damp	39	99 1-2	101 1.2	101 1.4	102	99 1 2 bef. rising
31	N E	Rainy	45	101 1 4	101 1.2	101 I.4		100 do do
Feb 1	N W	Clear	28	101			102	100 do do
Mar 26		do		100 1-2			101	
July 9	W	Cldy and damp		100				before rising
10	W	Clear	63	100	101			
11	N E	Cloudy	65	100	101			

2 M

Fig. 16: 'Table, showing the temperature of the interior of the stomach, in different conditions, taken in different seasons of the year, and at various times of the day, from 5 o'clock in the morning, till 12 o'clock at night'. Beaumont, *Observations* (1833), p. 273.

process, distinguishing digestion from rotting and decay. By the use of the aperture in the wall of St Martin's stomach Beaumont was able to show that digestion, as a process, was independent of whether it took place within the body or in a glass vessel, provided the temperature was comparable and the gastric juice present. By keeping the gastric juice sealed in a jar and trying it after a lapse of many years, Beaumont was able to show that it still had its old capacity to digest foods. Nor is it just an ancillary substance, merely moistening the food. It has quite specific digestive powers as van Helmont had supposed.

Summing up the results of years of patient study, Beaumont says, 'I think I am warranted, from the result of all the experiments, in saying that the gastric juice, so far from being "inert as water", as some authors assert, is the most general solvent in nature of alimentary matter – even the hardest bone cannot withstand its action. It is capable, *even out of the stomach*, of effecting perfect digestion, with the aid of due and uniform degree of heat (100° Fahrenheit) and gentle agitation. ... I am impelled by the weight of evidence ... to conclude that the change effected by it on the aliment, *is purely chemical.*'

By chance Beaumont was offered a kind of walking apparatus. But his work illustrates a further point about experiments. Logically his lengthy experiment exemplifies the intensive design very beautifully. Only one stomach was ever involved. Yet the scientific community never doubted that Beaumont's results applied to the stomachs of all mankind. Why? It can only be because no one questioned the principle that one stomach is very like another, and that which chance provides will do as an exemplar for them all (see below, p. 207).

Later work on the physiology of digestion

There was a kind of perfection about Beaumont's researches, so that he both opened and closed a chapter in the study of human physiology. Detailed investigations of the chemical reactions involved could not have been undertaken in his time. But there was an outstanding major problem in understanding the process of digestion left untouched by Beaumont's researches, though it was within the compass of nineteenth-century technique. How were the digestive ferments produced? Was the presence of the food material in the stomach enough to start them flowing? In 1889 Pavlov demonstrated conclusively that the stimulus that brought on secretion from the stomach was mediated by the nervous system. He operated on a dog to separate a small fold of the stomach lining communicating with the exterior through a fistula. Then he closed the aesophagus off and opened it to the exterior so that the food swallowed by the dog did not enter the stomach at all.

He showed that the moment the dog started eating the stomach secretions began and continued just so long as eating went on. Since no food entered the stomach the stimulus must have been mediated by the nervous system.

But it gradually became apparent that this mechanism would not account for secretions in parts of the digestive tract and associated organs other than the stomach. The role of hormones was first clearly established by W. M. Bayliss and E. H. Starling in 1902. They also used a dog as their experimental animal. By separating a part of the intestine, the jejunum, from the rest of the tract, they could stimulate it separately. They left the arterial and venous connections untouched but they cut all connections with the nervous system. When they put some dilute hydrochloric acid into the duodenum, which still remained fully connected to the digestive system, there was immediate pancreatic secretion. And when they did the same to the detached section of small intestine there was just the same effect. But there was no physical connection between this separated section and the rest, except via the blood vessels and the blood circulating therein. There must be a chemical agent secreted by the wall of jejunum when stimulated by the dilute acid, which is carried with the circulating blood to set the pancreas going. They called this substance 'secretin'. By taking samples from the wall of jejunum, and injecting them into the blood stream, they again produced the pancreatic secretion, which was not stimulated simply by injecting dilute acid.

Further reading

van Helmont, J. B., *Oriatrike or Physick Refined*, transl. J. Chandler, London, 1662.

Beaumont, W., *Experiments and Observations on the Gastric Juice and the Physiology of Digestion*, Plattsburg, Va., 1833; Edinburgh, 1838.

Myer, J. S., *Life and Letters of Dr William Beaumont*, St Louis, 1912; 2nd edn., 1939.

Rosen, G., *The Reception of William Beaumont's Discovery in Europe*, New York, 1942.

B

Deciding between
Rival Hypotheses

The simplest logical structure within which a deliberately contrived experiment can be effective is that in which a single hypothesis entails a testable prediction, against a background of relatively fixed and stable theory and ancillary hypotheses. But it is almost, if not quite, impossible to find an example of an experiment which illustrates such a simple format. In real science hypotheses are usually tested in pairs, the one conceived as a rival to the other. The three experiments cited in this section were undertaken as ways of deciding between competing hypotheses, by testing consequences. **Robert Norman** set about trying to decide whether the tendency of magnetized needles to point to the geographic north was the result of an attraction from some northern point, or whether the whole magnet was orienting to some structured property of some kind of primitively conceived field. Among **Stephen Hales**'s many experiments was an elegant test of rival hypotheses about the movement of sap in plants. Did it circulate like the blood of animals, or did it flow in a more or less tidal way? When **Konrad Lorenz** was trying to find out the details of the process by which the young of a species become 'imprinted' with suitable adults, he needed to find a test for whether all the necessary behavioural routines were involved in a single act of imprinting or whether the imprinting of appropriate targets occurred separately.

But the truth of a consequence does not prove the truth of the hypothesis from which it follows, though the rival is eliminated as false. Successful experiments in this mode still leave open the possibility of further revision. This point is illustrated particularly in the work of Norman, and the subsequent history of the hypothesis he thought he had established.

3. Robert Norman

The Discovery of Dip and the Field Concept

Robert Norman was born about 1550. Nothing is known of his early life or family. He spent some 18 to 20 years at sea, as a navigator. It seems likely that he lived for some of that time in Seville. We know of him first through his work as an instrument-maker for William Burroughs. As a practical sailor Norman was well aware of the shortcomings of the navigational techniques and instruments of his day. The magnetic compass had become the most important navigational instrument, and Norman's discoveries were centred round its development for sea-going use. The variation of the magnetic north from the true bearing was well known and had been supposed to be a systematic and regular effect, that could be used in determining longitude. But by years of questioning of sailors, particularly traders on the 'Muscovy' route, he was able to establish that the proportional theory of variation was false. Then he discovered 'dip', the tendency of the magnetized needle not only to turn towards the north but to swing down from the horizontal in a regular fashion. He called this phenomenon 'declination'. Norman suspected that dip would be proportional to the latitude at which it was measured, and that an instrument could be devised to exploit this possibility. To this end he developed the dip-circle, a needle mounted on a horizontal pivot moving against a vertical graduated circle. He brought out his magnetic discoveries in *The Newe Attractive*, published by Ballard in London in 1581.

Norman was given to poesy of a sort and begins the book with a verse or two in praise of the magnetic effect. It takes the form of a challenge from the useful Lodestone to the merely decorated gem stones.

> Magnes, the Lodestone I,
>   your painted sheaths defy,
> Without my help in Indian sea,
>   the best of you might lie.

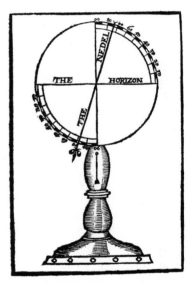

Fig. 17: The dip circle. Diagram from Norman, *The Newe Attractive* (1581), p. 10.

And several other verses to like effect.

In 1590 he published *The Safegarde of Saylers*, a translation of a Dutch navigational manual for the sea crossings from continental Europe. This was the first book in English to include woodcuts of the appearance of the coast from the sea. It too includes a poem, 'in commendation of the painful seamen'.

If Pilot's painful toil be lifted then aloft
for using of his Art according to his kind,
what is due to them who first this Art outsought,
And first instructions gave to them that were but blind?

Norman lived in a house in Radcliffe, close to London, from which he sold instruments for navigation. Little is known of his personal circumstances and one can only conjecture that he must have died somewhere about 1600, the date of the

Fig. 18: Title page of *The Newe Attractive*, in a reprint published as late as 1720.

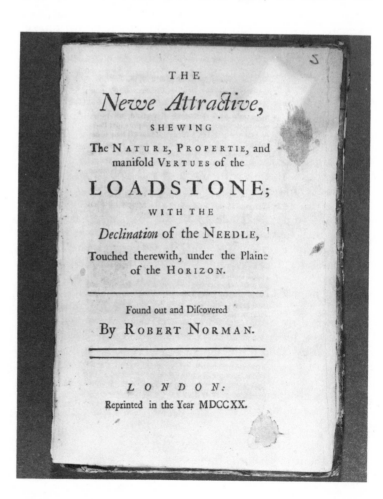

publication of Gilbert's *De Magnete*, a work in which Norman's discoveries were much advanced.

An irritating anomaly: the discovery of dip

The experiment to be described in this section was the first step towards the realization of the idea of a magnetic *field*. But as in many of the studies we shall be examining, the central and most illuminating experiment was part of a series of discoveries, an exploration of a family of phenomena. In this case the research programme was sparked off by a quite small anomaly.

Norman gives a vivid description of the occasion on which he discovered dip. He noticed that even with his most carefully constructed compasses the magnetized needle, when balanced on a smooth pivot, would not only turn to the north, but that the north end would decline, or as we should now say, dip. This effect had to be compensated for in the construction. He was '... constrained to put some small piece of ware in the south part thereof, to counterpoise this declining, and to make it equal again'. But he had not considered making an independent study of a tiresome but peripheral effect. One day, however, after having made a very fine needle and pivot, he found the declination was very strong, so he began to cut the needle, to shorten the north segment. '... in the end', he says, 'I cut it too short, and so spoiled the needle wherein I had taken such pains. Hereby being strocken into some choler, I applied myself to seek further into this effect.'

The first step was to construct a dip circle, so that systematic measurements of the effect could be made. By pivoting the needle on a horizontal axis the full effect could be produced, and its extent measured.

But was it due to magnetization, or to some side effect produced by the lodestone? The most obvious possibility was that the north end had taken up some 'ponderous or weighty matter' from the lodestone. Norman devised a simple test of this idea. He put some small pieces of iron in a balance pan and made up an equal counterweight of lead, which is non-magnetic. Then he magnetized the iron and the result was clear. 'You shall find them to weigh no more, than before they were touched. Furthermore if the north end of the needle had taken up something weighty from the lodestone, so too 'the south end should have taken up something weighty from the other end of the lodestone, and there would be no dip effect.' Two questions had to be settled: 'By what means this declining or elevating happeneth', and 'In which of the two points [north pole or south pole] consisteth the action or cause thereof?'

It had been assumed by Norman's predecessors that the tendency of the magnetic needle to swing towards the poles was due to a 'point attractive' that drew the north-seeking pole. But 'if we can show there is no attractive or drawing power then there is no point attractive.' But the needle does turn towards a point. This should then be called the 'point respective'. Just a name, one might say. But the choice of name carries with it the weight of theory. If the point marks a source of attraction, then one would expect a force acting between the pole of the magnetic needle and that source, drawing the needle. But if the point marks some structural property of the medium, no such drawing force is to be expected.

Proving the field concept

The experiment devised by Norman to settle the question is of the greatest elegance. (And as we shall see did not settle the matter from our present viewpoint.) 'Now to prove no Attractive point ... you shall take a piece of iron or steel wire of two inches long or more, and thrust it into a piece of close cork, as big as you think may sufficiently bear the wire on the water, so as the same cork rest in the middle of the water.

'Then you shall take a deep glass, bowl, cup or other vessel, and fill it with fair water, setting it in some place where it may rest quiet, and out of the wind. This done, cut the cork circumspectly by little and little, until the wire with the cork be so fitted, that it may remain under the superficies of the water two or three inches, both ends of the wire lying level with the superficies of the water, without ascending or descending, like to the beam of a balance being equally poised at both ends.

'Then take out the same wire, without moving the cork, and touch it with the stone, the one end with the south of the stone, and the other end with the north, and then set it again in the water, and you shall see it presently turn itself upon its own centre, showing the aforementioned declining property, without descending to the bottom, as by reason it should, if there were any attraction downwards, the lower part of the water being nearer that point, than the superficies thereof.'

It seems that there is no pulling or drawing of the whole needle from its north end by attraction from some northerly point in the earth or the heavens. We must, thought Norman, attribute the whole power to point to the north 'to be in the Stone, and in the needle, by the virtue received of the stone.' With hindsight we know that there was another hypothesis that neither Norman nor Gilbert had thought of, that there were both attractive and repulsive forces which depended for their strength on the distance of the sources. Thus the needle would turn to the north pole of the earth because of a balance between forces of attraction and of repulsion, between both poles of the magnet and both poles of the earth. Happily this more complex force theory, the central focus of argument between Ampère

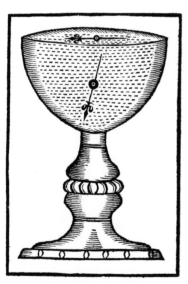

Fig. 19: The 'wine glass' experiment, showing a magnetized needle suspended in water. Diagram from *The Newe Attractive* (1581), p. 14.

and Faraday some 250 years later, occurred to neither of the great Renaissance students of magnetism. Norman and Gilbert after him dealt with the problem of explaining terrestrial magnetism by the invention of the idea of the field of force, the foundation idea of the modern physics of electricity, magnetism and gravity.

Norman says, 'And surely, I am of the opinion, that if this virtue [magnetic power] could by any means be made visible to the eye of man, it would be found in spherical form extending round about the stone in great compass, and the dead body of the stone [lies] in the middle thereof, whose centre is the centre of his aforesaid virtue.'

This idea is prescient but radically incomplete. Norman ascribed a magnetic field only to the lodestone. He says nothing about the earth. Gilbert made the final step, in his *De Magnete* of 1600. He repeated Norman's experiment, to demonstrate that neither dip nor the tendency of the needle to seek the north were to be explained by attraction (so he thought). But just ascribing a field to the lodestone and the needle is not enough. The earth too is a magnet and must therefore have (or perhaps even ultimately be) a magnetic field. This Gilbert called the *orbis virtutis*, the sphere of power. His conclusion from Norman's experiment goes further: 'Yet the direction is not produced by attraction but by a disposing and conversory power existing in the earth as a whole.' It is the *orbis virtutis*

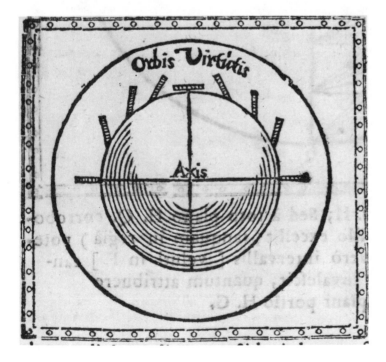

Fig. 20: Gilbert's 'orbis virtutis' or sphere of power. Diagram from the *De Magnete*, 2nd edn, Stettin (1628), p. 78.

Robert Norman

that is responsible for the setting of the needle in a specific direction.

By making a model earth or 'terrella' out of a spherical lodestone Gilbert was able to show in miniature how dip would vary with latitude. This was a much more promising navigational idea than Norman's, since he had not formulated the idea of the earth's magnetic field.

Here in Gilbert's own words is the moment of birth of the true field conception. '... such is the property of magnetic spheres that their force is poured forth and diffused beyond

Fig. 21: Gilbert at work. *De Magnete* (1628), plate xi.

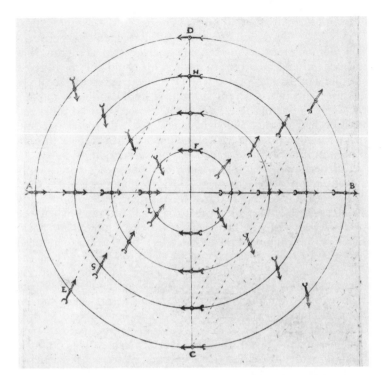

Fig. 22: Gilbert's conception of the
structure of a magnetic field. *De
Magnete* (1628), plate xii.

their superficies spherically, the form being exalted above the
bounds of corporeal matter. . . . magnetic bodies do not regard
the same part or point of the terrella at every distance whatever
therefrom, . . . but ever do tend towards those points of the
spheres of influence which are equal arcs distant from their
common axis . . . we do not mean that the magnetic forms and
spheres exist in the air, or water, or any other medium not
magnetical . . . in the several spheres magnetic bodies control
other bodies magnetical and excite them even as though the
spheres of influence were solid, material lodestones.'

Subsequent developments: the re-invention of the magnetic field

But two steps remained to be taken. Magnetic and electrical
studies were relatively neglected for some 150 years. How to
achieve Norman's dream and render the spheres of influence
and power visible? Now any schoolboy knows that you must
just shake a few iron filings on a sheet of paper under which
there is a magnet. Immediately the *orbis virtutis* and its lines of
force which are 'exalted above the bounds of corporeal matter'
become visible. This idea and the subsequent experimental
investigations of the properties of the lines of force we owe to
Faraday.

Both Norman and Gilbert had conceived the magnetic field to be independent of matter. Faraday succeeded in experimentally demonstrating the fact. He showed that a rotating bar magnet induced a current in itself. This could only be because the metal of the bar rotated while the field, represented by lines of force, did not. Current is induced by a conductor moving relative to a line of force, 'cutting it' as we say. So if the metal magnet had carried the field round with it, metal bar and magnet field would have been stationary with respect to each other, and there would have been no current.

Furthermore, Faraday had demonstrated that merely switching on an electromagnet and switching it off would induce currents. It seemed that the magnetic field produced by an electrified wire took time to spread out, and when the powering current was turned off, it again took time to fall back. Faraday had demonstrated this effect by detecting induced currents in wires placed close to electromagnets. When the current was off, and also when it had been on for a time, no current was induced in the wire. But when the electromagnet was switched on and *after* it was switched off there was an induced current. These and other effects convinced Faraday, and I suppose they serve to convince most of us, that fields are real, part of the furniture of the world. Perhaps it is only the limitation of our senses that prevents us from experiencing fields in as a direct a way as we are aware of earth and water.

The wine-glass experiment fills out the theory of experiments a little more. Theodoric and Aristotle carried out observational and experimental studies that bore positively on their results. Norman's experiment depends on a more complex logic. He conceived the result as a *refutation* of the attraction hypothesis and an *illustration* of the field concept. In this example we add our second and third aspects of experimental science to the positive or inductive aspect. And since the effect that is supposed to illustrate the field concept is explicable in terms of another, more sophisticated version of the attraction theory, we can take heed too, of the danger of supposing that every explanation that works must be the true account of the causes of a phenomenon.

Further reading

Norman, R., *The Newe Attractive*, London, 1581.
Gilbert, W., *De Magnete*, London, 1600.

Roller, D. H. D., *The De Magnete of William Gilbert*, Amsterdam, 1959.
Waters, D. W., *The Art of Navigation in England in Elizabethan and Early Stuart Times*, London, 1958.

4. Stephen Hales

The Circulation of Sap in Plants

Stephen Hales was born to a well-to-do family in Bekesbourne, Kent, in 1677. In 1696 he entered Bene't College, Cambridge. At that time the educational opportunities at Cambridge were remarkably diverse. With his friend William Stukeley he seems to have combined extensive studies in natural history and biology with a great interest in the physics of fluids, gases and liquids. So from the very earliest knowledge we have of him, the *Leitmotif* of his scientific and engineering work was apparent, the role of pneumatic and fluid dynamics in the processes of life.

Hales remained in Cambridge as a Fellow of his College until 1709, when he became the Vicar of Teddington, a post he held for the rest of his life. Though Harvey is credited with the 'discovery' of the circulation of the blood in men and animals, this amounted to no more than a theoretical demonstration of the necessity of such a hypothesis given the facts about how much blood the body contained. In a long series of both gruesome and rigorous experiments on horses, dogs and frogs, Hales explored many aspects of the blood vascular system, charting its pathways and exploring the hydrodynamic conditions of pressure and flow that characterized each part. His work was definitive, solving many of the major problems left by Harvey's inspired hypothesis.* But these investigations did not pass unnoticed by the general public. Thomas Twining

Fig. 23: Stephen Hales, aged 82. Engraving by T. M. Ardell after a painting by T. Hudson. Museum of the History of Science, Oxford University.

---

* 'A broad concept of blood pressure, blood flow, blood velocity and their relations, and quantitative measurements or calculations of each – these were the great contributions of Stephen Hales to the knowledge of the output of the heart, a contribution which has oriented all future work.' W. F. Hamilton and D. W. Richards in *Circulation of the Blood: Men and Ideas*, edited by A. D. Fishman and D. W. Richards, New York, 1964.

includes a verse in his topographical poem *The Boat* that runs
as follows:

> Green Teddington's serene retreat
> For philosophic studies meet,
> Where the good Pastor Stephen Hales
> Weighed moisture in a pair of scales,
> To lingering death put Mares and Dogs,
> And stripped the Skins from living Frogs.
> Nature, he loved, her Works intent
> To search, and sometimes to torment.

Though the movement against thoughtless cruelty to animals
had begun about this time, its leading protagonist, Alexander
Pope, a neighbour of Hales, became one of his closest friends.

In about 1724 Hales began the series of studies that
established the main outlines of the physiology of plants. Not
only did he study the way the sap circulated, but most
importantly the interactions and exchanges between the plant
and its environment. He showed how the water drawn in by
the roots is transported to the leaves and there transpired.
Growth, too, interested him, and he demonstrated how the
various parts of plants grow and in what proportions. Mayow
had shown the relation between respiration, combustion and
the air some years before Hales began his studies, and he
pursued this problem too.

In 1722 Hales was elected a Fellow of the Royal Society,
becoming a member of the Council in 1727. He had become a
public figure of considerable eminence, a Trustee for the
Colony of Georgia, and a regular member of commissions
appointed to look into matters of public health, such as
conditions in the ships of the Royal Navy, and the examination
of alleged wonder cures. His interest in air extended to
ventilation. The problem of getting fresh air into confined
spaces such as the living quarters of ships, prisons and
hospitals became for a while his chief preoccupation. He
invented a variety of ventilating devices, most of which were
put into practical use. He died in 1761, still the Vicar of
Teddington.

Early work on the hydrodynamics of plants

Botanical studies in antiquity were dominated by the work of
Theophrastus, a pupil of Aristotle. Most works were descriptive and classificatory, grouping plants by reference to their
general form, such as herbs, bushes and trees, or by their
alleged medicinal properties. These classifications came
through into the Middle Ages. They were thoroughly practical
in intent, if greatly corrupted in substance after innumerable
and inaccurate copyings. Theophrastus had also made some

study of the relation of plants to their typical environments, cross-classifying by reference to habitat. But this aspect of his work had degenerated into little more than a guide for where to look for specific herbal remedies. So far as we know there were no anatomical or physiological studies of plants in antiquity.

The first substantial modern work was made possible by the development of the microscope in the mid-seventeenth century. Robert Hooke, the same who had served as Boyle's assistant, made careful microscopical examinations of plants. He was the first to identify the cell as the basic biological unit. Nehemiah Grew carried this kind of work very much farther, making detailed studies of the anatomy of plants, and producing anatomical drawings of the highest quality.

The Circulation of Sap in Plants

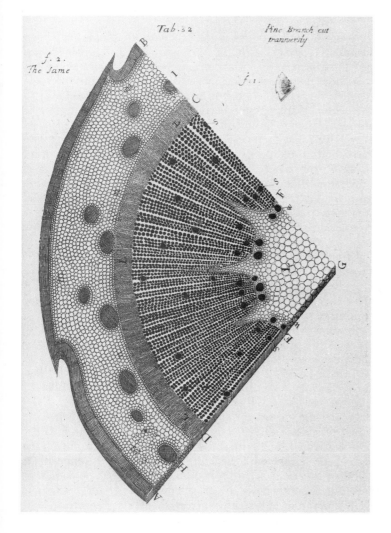

Fig. 24: Anatomical structure of a pine branch, from N. Grew's *The Anatomy of Plants*, 2nd edn, London (1682), table 32.

The most important discovery to come out of the use of the microscope was the realization that the plant contained ramifying systems of tubes, running from the roots through the stem and branches to the leaves. Some of the tubes seemed to be filled with liquid, others with air. Considering this system Grew came to realize the possibility of a circulation in plants comparable to that known to occur in animals. Once this thought was formulated all sorts of questions sprang to mind. Was there a *closed* circulation in plants as there was in animals? What force powered the flow of sap? Relative to this circulation what were the life functions of the various parts of the plant? It was to these questions that Hales devoted his great experimental series.

The circulation of the sap

The basic theory of the vital processes of plants had been formulated about 1670 by Malpighi. He had grasped two points of crucial importance. Common sense had suggested that there must be a movement of sap upwards from the roots towards the leaves, contributing at least the watery element to the whole plant. Malpighi realized that the elaboration of simpler elements into plant substance took place in the leaves. It followed that there must also be a downward movement to carry body-building substances from the leaves to the other parts of the plant where they were to be used. He also understood the process that led to the production and storage of a surplus for later use. Since in many plants this material was stored in tubers associated with the roots, the counter-circulation of nutriments must reach as far as the roots, the very source of the primary circulation of water. All this was informed speculation. It remained to be demonstrated experimentally. This was Hales's contribution.

As in so much scientific work the central experiment which I shall describe was the culmination of a series of subsidiary experiments preparing the way for it. First it was necessary to determine whether the throughput of water from roots to leaves was a process powered by pressure from the roots or by some drawing process from the leaves.

'July 27 [1716]. I fixed an *Apple-branch* ... to a tube. I filled the tube with water, and then immersed the whole branch ... into the vessel *u u* full of water.'

'The water subsided 6 inches the first two hours (being the filling of the sap vessels) and 6 inches the following night. ... The third day in the morning, I took the branch out of the water; and hung it with the Tube affixed to it in the open air; it imbibed this 27 + ½ inches in 12 hours.' Hales concluded that this experiment 'shews the great power of perspiration'. It is the evaporation of water from the leaves, not the pressure of

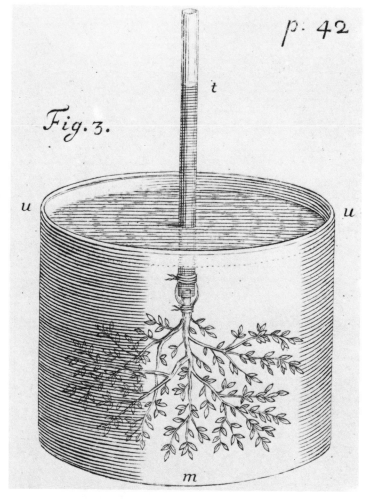

Fig. 25: This plate from Hales's
Vegetable Staticks illustrates the
experiments to show that evaporation
from the leaves is involved in the ascent
of sap. 3rd edn, London (1738), plate
2.

water in the roots that is the prime mover in the circulation of
the sap. Of course these experiments do not show how these
processes come about.

But is it water that is transpired from the leaves? That the
fluid is mostly water can be demonstrated neatly by confining a
leafy branch in a vessel and collecting the 'perspired' fluid.

Now the stage was set for the key experiment: how does the
sap move? Is it a circulation as the animal analogy would
suggest, or is it a kind of tidal ebb and flow? In two perfect
experiments Hales cleared this matter up for all time. The
circulationists had assumed that the sap moved up in the inner
part of the stem and down in the outer.

On August 20 [1716] he says, 'at 1 *p.m.* I took an *Apple-
branch b* nine feet long, $1 + \frac{3}{4}$ inch diameter, with proportional lateral branches, I cemented it fast to the tube *a*, by

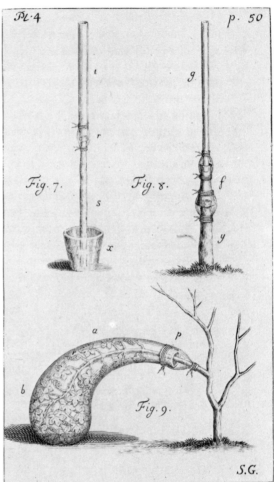

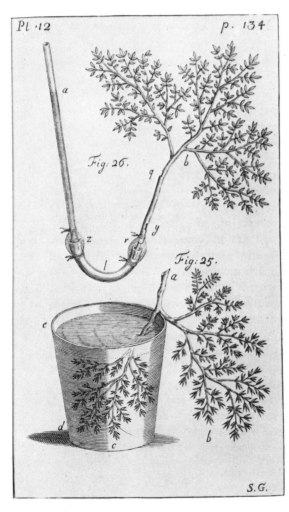

Above Fig. 26: Another plate from the *Vegetable Staticks* (1738). Fig. 9 illustrates the collection of 'perspired' liquid from the leaves. It was water.

Left Fig. 27: In this plate from the *Vegetable Staticks* (1738), Fig. 26 Illustrates the third experiment. The notches used to test the 'circulation' hypothesis can be seen at *y* and *q*.

means of the lead syphon *l*; but first I cut away the bark, and last year's ringlet of wood, for 3 inches length to *r*. I then filled the tube with water, which was 22 feet long and ½ inch diameter, having first cut a gap at *y* through the bark, and last year's wood 12 inches from the lower end of the stem: the water was very freely imbibed, *viz.* at the rate of 3 + ½ inch in a minute. In half an hour's time I could plainly perceive the lower part of the gap *y* to be moister than before; when at the same time the upper part of the wound looked white and dry.'

It follows that 'the water must necessarily ascend from the tube, through the innermost wood, because the last year's wood was cut away, for 3 inches length all round the stem; and consequently, if the sap in its natural course descended by the last year's ringlet of wood, and between that and the bark (as many have thought) the water should have descended by the last year's wood, or the bark, and so have first moistened the upper part of the gap *y*; but on the contrary, the lower part was moistened, and not the upper part.' Since the sap must be ascending by the inner part of the stem, there being a ring cut right round below the gap *y*, and since it is also ascending by the last year's wood and the bark, as evidenced by the moisture forming at lower side of the gap, there is no circulation, at least not in the strict sense of a complete hydraulic cycle. If there had been a cycle, movement in one direction in one part would have been compensated for by correlative movement in another direction, somewhere else in the system.

Further strong indirect evidence can be found for this conclusion, from a consideration of how much water a plant takes up and transpires in a day. Hales showed that the sunflower transpires water at a rate seventeen times that of a man, bulk for bulk. If there were a circulation it would have to be enormously fast. But there is no evidence whatsoever for such celerity of movement.

But 'the sap does in some measure recede from the top to the bottom of plants', as many ingenious experiments have proved, so Hales notes. But this does not demonstrate a circulation, rather a daily ebb and flow.

Developments in plant physiology after Hales

It is quite fair to say that in the hundred years immediately following the masterly series of experiments of which I have described only one particularly ingenious fragment, Hales's successors added little to the science of plant physiology. However, some contributions were made in this period. Hales's experiments had almost fully clarified the water economy of plants. But plants are also exchanging gases with the atmosphere. Mayow (the first scientist clearly to distinguish the gases of the atmosphere) and Hales had both suspected that plants took some of their nourishment from the air. Hales had distinguished the kinds of gaseous exchange, the nutritive and the respiratory. But he had failed to understand Mayow's discovery of a *constituent* of air, 'spiritus nitro-aereus' (or 'oxygen' as we now call it), which was absorbed or 'fixed' in vital processes. Hales had supposed that respiration and combustion reduced the volume of air by one fifth because the air had lost that proportion of its elasticity, rather than that one fifth of its substance had been absorbed. Given this quite central error in his theory of the air Hales was unable clearly to identify the nutritive and respiratory gaseous exchanges for what they were. In 1779, the Dutch doctor Ingenhousz established that there were two quite distinct respiratory cycles in the life of plants. In one cycle oxygen was absorbed and carbon dioxide exhaled just as in animal respiration. In the other cycle carbon dioxide was taken in as a kind of gaseous food, and oxygen was given out. By about 1840 the chemistry of the gases of the air was well known. Oxygen, nitrogen and carbon dioxide had been clearly distinguished and their chemical properties thoroughly investigated. The final step came in 1840 when Boussingault showed that plants obtained their nitrogen not from the air, but from the nitrates present in the soil in which they grew.

Further reading

Hales, S., *Vegetable Staticks*, London, 1727. A fine modern reprint has been edited by M. A. Hoskin, Oldbourne Science Library, London, 1961.

Allan, D. G. C., and Schofield, R. E., *Stephen Hales: Scientist and Philanthropist*, London, 1980.

Clark-Kennedy, A. E., *Stephen Hales, D.D., F.R.S.: An Eighteenth Century Biography*, Cambridge–New York, 1929; repr. Ridgewood, N.J., 1965.

von Sachs, J., *History of Botany*, transl. H. E. F. Garnsey and I. B. Balfour, Oxford, 1906.

5. Konrad Lorenz

The Conditions of Imprinting

Konrad Lorenz was born on 7 November 1903. He was the second son of Adolf Lorenz, an orthopaedic surgeon of great skill and enormous international reputation. Adolf Lorenz had developed a successful treatment for congenital hip dislocation and became extremely rich through the cultivation of an international practice. Konrad Lorenz's childhood was spent mostly in the vast house his father built in the village of Altenberg, close to the river Danube and not far from Vienna. As a child he kept all kinds of animals, ducklings, fish, dogs, and built up a colony of Jackdaws in the attics of the house, an avian society that provided him with the material of his first scientific paper. From the age of eleven he attended the Schottengymnasium in Vienna, and when the difficulties of transport from the village into town became acute in the First World War, the family moved into a flat in the city.

Fig. 28: Konrad Lorenz

Adolf Lorenz was keen for Konrad to follow him into the medical profession, and sent him off to New York to take the premedical course at Columbia University in 1922. Young Lorenz disliked this and very soon returned home. He then entered the Medical Faculty at the University in Vienna, but to study anatomy as a science rather than to proceed to a medical career. At this time he was much influenced by a close friend, Bernard Hellman, who shared his interest in natural history. Lorenz published his first paper, 'Observations on Jackdaws', in 1927, and a year later took his doctorate in medicine.

Instead of starting out in medical practice he became an assistant in the anatomy department. At this time he made the acquaintance of the first systematic student of natural animal behaviour, Oskar Heinroth. It is clear from the many references Lorenz makes to Heinroth that he learned a great deal from him. Lorenz took a doctorate in zoology in 1933, and moved to that department. His basic scientific work was done in the years 1926 to 1938. Though he has continued active research to this day his great discoveries were made in those twelve years.

Lorenz had always been greatly interested in the River Danube. During the 1930s he bought a boat, and took the trouble to take the Danube Riverboat Pilot's examination. In 1930, he married Margarethe Gebhardt, whom he had known all his life.

The Second World War totally disrupted his scientific work. His biographer, Alec Nisbett, reports him as being rather naive politically, not fully awake to the nature of the Nazi regime until well into the war. His medical qualifications drew him in as an Army doctor, and he served in Poland from 1941. From there he moved to the Eastern Front, and was eventually captured by the Russians in 1944, spending altogether three years in captivity, mostly in Soviet Armenia.

After the war the development of scientific research in Germany and Austria was much hampered by the controls imposed by the occupying powers. Eventually the Max Planck Institute was formed in Göttingen in 1948. The Society which governed the Institute was prepared to support Lorenz's work. He used his own home in Altenberg as a combined Institute and field station. In 1951, thanks to the assistance of Baron von Romberg, a Max Planck Institute specifically devoted to ethological research was set up in Balder, and eventually at Seewiessen. Lorenz became the Director in 1962.

In 1974 he shared the Nobel Prize with Niko Tinbergen and Otto von Frisch.

Early work in ethology

The study of animals in their natural environment, leading their ordinary lives, had long been the province of natural historians, usually amateurs. Only Darwin had given the study of natural animal behaviour a scientific turn. He had seen the central idea of ethology, that animal behavioural routines should be regarded as aspects of the animal's adaptation to its environment quite as important as its anatomical structure or its physiological processes. And he had drawn the conclusion that routines must be inherited and naturally selected. There the matter rested more or less. When the study of animal behaviour again began to interest scientists it was in the United States. The prime originator of the idea that animals can be understood only through a prolonged acquaintance with their normal lives in a natural environment was the American biologist, C. O. Whitman. He advocated a Darwinian approach to the explanation of behavioural routines. Whitman and his students, amongst whom was the influential W. M. Wheeler, were doing work of the highest quality on the natural behaviour of very diverse species. Lorenz himself has said that his greatest achievement was to have brought together the work of Whitman and that of his own mentor, Oskar Heinroth.

But the original insights of Darwin were neglected by most psychologists. Animals, particularly primates and rats, were subjected to endless experiments in caged conditions, to try to discover the elementary units of behaviour and the stimuli that elicited them, and the process by which the supposed elementary reactions had been conditioned. The entire programme was radically misconceived, but it had become so entrenched that progress could only come by researches that developed independently of it.

The transformation of the study of animal behaviour came through the application of rigorous standards to natural observations of the life forms of animals living in their ordinary environments. The beginnings of this new animal science, ethology, were in Germany. Very soon, however, this work was most fruitfully brought together with a native British tradition of natural history and naturalistic observation of the habits of animals in the wild. But the key figure at the centre of the new field was Lorenz.

The Conditions of Imprinting

Fig. 29: A laboratory experiment involving monkeys in a highly artificial situation.

Konrad Lorenz

If Darwin was right, then there would be naturally selected *routines*, elaborate, integrated chains of behaviour directed to the achievement of certain ends adaptive to the breeding success of a species. The first ethological studies were concerned with the identification of these routines, the demonstration that they could not have been learned, and the working out of the details of the ways that this or that routine, integrated with other routines, was adaptive to successful breeding. This can be illustrated by the case of the routines of eggshell removal from nests as young birds hatch out. A newly opened eggshell shows a bright white interior, easily proved to be attractive to predators. Most birds which nest in places exposed to predators have an innate routine of eggshell removal. But those which nest in remote places, safe from predation, do not inherit the neurological basis of any such routines.

But there is more to activating a routine than merely inheriting the neurological machinery that operates the chain of reflexes for running through the action sequence. The routine must be triggered by the right kind of stimulus. The question now arises: are the capacities to recognize the right stimulus inherited along with the capacity to play through the routine when stimulated? It turns out that the answer is disconcerting, 'only sometimes'. The young of many species of bird do not recognize conspecifics, birds of their own kind, if they have not been introduced to them at a definite period in their development.

The young of godwits, for example, are hatched at an advanced stage of development and have an 'innate schema', by which they recognize the adult bird so that they immediately display appropriate behaviour in the presence of adults, say gaping for food. They flee from human beings without any special prompting or learning. Experiments have shown which adult characters are important. By imitating each adult characteristic separately, the appropriate reactions of young birds can be identified. For these birds and similar species the capacity to recognize the object of a behavioural routine, as well as the capacity to perform the routine, must be innate or inherited.

However, most birds develop quite differently. The Greylag Goose has become famous in ethological circles as the species most vividly displaying another pattern of development. When goslings are reared wholly by human beings it is towards humans that the young geese direct their behavioural routines. They seem to acquire as a prime object of interest whatever creature happens to be present at the right moment in their development. The first recorded observation of the phenomenon (now called 'imprinting') is due to Oskar Heinroth. He noticed that though ducklings rush away and hide from humans directly they are hatched, goslings 'stare calmly at human beings and do not resist handling... The young gosling [so treated] shows no inclination to regard [adult geese] as conspecifics ... it regards the human being as a parent.' As Heinroth remarks, freshly hatched goslings stare out from the debris of their shells 'with the intention of exactly imprinting ... [the first things they see]' as the image of their parent.

Fig. 30: Lorenz being followed by three hand-reared imprinted Greylag Geese.

70

Konrad Lorenz

Fig. 31: Drawings made by Lorenz to illustrate his own book, *King Solomon's Ring*.

The experimental discovery of the timing conditions of imprinting

Lorenz's contribution was the systematic experimental exploration of the conditions of the occurrence of this phenomenon. His first discoveries sharpened up the basic idea of imprinting. By comparing several species he demonstrated that the 'object can only be imprinted during a quite definite period in the bird's life'. After the imprinting has occurred and that period has elapsed (and the length of it varies greatly with different species), 'the recognition response cannot be forgotten'. Two very important theoretical conclusions follow: contrary to the usual assumption of conditioning in animal studies, there must be an innate drive to 'fill this gap [lack of a specific object to which to direct the behaviour] in the instinctive framework.' But even more importantly, it would be quite wrong to think of imprinting as a kind of learning. It is characteristic of learnt routines that they can be forgotten or displaced by other learning. But once a creature has been imprinted on a particular species as the target for some instinctive patterns of behaviour, 'the animals that have been imprinted do not alter their behaviour in the slightest' while the appropriate behaviour pattern is part of their life requirements, such as gaping to be fed.

The central experiment to be described in this section was designed to determine whether all instinctive behavioural routines were directed to one and only one type of imprinted object because the object for all of them was imprinted together, or whether each routine had, as it were, its own imprinting moment. If the latter were the case each routine could have as its object a distinct individual, each from a different species, if it were around at the crucial imprinting time for that routine. The object of the study was one young jackdaw from Lorenz's extensive colony, living in the attics of the house at Altenberg. The bird had been reared in complete isolation from jackdaws and other birds, so that all but two of its normal repertoire of instinctive behavioural routines were either innate or had been imprinted on human beings. Of these two, one, the routine of flying in the company of a flock, had been imprinted on hooded crows, these being the first birds of the right type with which the jackdaw had become acquainted during that period of its life in which appropriate objects for companionable flying could be imprinted. Even when well grown and living in the company of other jackdaws, it flew off every day to join a flock of hooded crows and spent its time with them. This established the independence of at least one routine with its appropriate object and moment of imprinting. But the jackdaw was living amongst other jackdaws when the critical period of imprinting the objects of reproductive routines occurred. So it directed its mating advances to other jackdaws. It mated with jackdaws, but flew with hooded crows, and fed with people. The imprinting of the reproductive routines and the imprinting of the flying routines must have occurred on separate occasions in the life of the jackdaw. Normal jackdaws fly, mate and feed with other jackdaws, but the experiment suggests that the objects for each major life routine were imprinted at different times. There must therefore be a programmed sequence of moments at each of which imprinting for a specific routine must occur. But there remained the routine associated with the care of the young. When the jackdaw of the experiment first came across a fledgling jackdaw, 'abruptly', says Lorenz, it adopted the young bird and 'guided and fed it in a completely species specific manner'. But this was the first fledgling jackdaw that it

Fig. 32: Lorenz's family home at Altenburg.

Konrad Lorenz

had seen, so there could not have been a prior imprinting of the object of that routine. One must conclude then that not only are there specific moments for routines that require imprinting of appropriate objects to be complete, but there are also, in the same species, routines which are innate with respect both to routine and to object.

One final point of principle remains. When a young bird is imprinted on an appropriate object, this object is a representative of a species. Does the bird imprint on the species, or on the object as an individual? The answer is somewhat complex. Lorenz found that if a bird had acquired a human being as a surrogate parent by imprinting, and continued to live with human beings as its sexual instincts were developing, it would direct these not at the one on which it had been imprinted as a parent, but on another human being. The innate mechanism controlling the imprinting process must be relatively complicated. At the earliest moment when it acquires a parent, so to speak, it selected *whatever* happened to be around, be it human or bird. But at a later stage when a mate is adopted, the imprinting process seems to fix the image of a particular individual.

Fig. 33: Lorenz with some of his family of Jackdaws at Altenburg in the 1930s.

N. Tinbergen, who shared the Nobel Prize with Lorenz and von Frisch, has continued the naturalistic study of behaviour patterns of a wide variety of creatures, and related these, more closely than Lorenz had done, to the neurophysiological aspects of the behaviour. Macfarland, a former pupil of Tinbergen, has carried this kind of study a stage further, by applying the concepts and methods of system theory to the formulation of hypotheses about the neural mechanisms that produce the pattern behaviour. But as Tinbergen has insisted, the persistence of patterns must be seen in a Darwinian framework, that is, organized behaviour should be thought of as adaptive to mating success of individuals relative to their natural environment.

Political assumptions as deep as those that lay behind the insistence that learning was the source of behavioural routines, the tacit belief that dominated early laboratory work on animal behaviour, are not easily set aside. Ethologists of the Anglo-European tradition have been persistently driven to defend their innateness hypothesis. This has led to great theoretical refinement and a wide range of observational and experimental studies to test the theory. It seems fair to say that at this stage there can be no serious doubt that the basic ideas of Lorenz and Tinbergen have stood the test of time.

In recent years the naturalistic method of studying animals in their ordinary habitats in an endeavour to understand the way they live out their lives has been extended to primates, and in particular to chimpanzees. There have also been very detailed studies of lions, gorillas and other large animals.

Along with progress in the scientific analysis and understanding of the lives of animals a flourishing secondary literature has grown up, devoted to drawing out comparisons between animal and human life. Most of the semi-popular works in this genre have tended to suggest that human beings too are innately programmed to perform certain sorts of routines. It has even been proposed that there might be some kind of imprinting of appropriate objects in human infants. The arguments of ethological popularizers (for instance, Robert Ardrey) have usually taken the form of speculative analogies between aspects of contemporary human behaviour and some of the behavioural routines of animals. These speculations have depended on imaginative reconstructions of the remote past of the human race, from whence its present habits are supposed to have descended.

Lorenz did not discover imprinting. But his experiments and observations decided between two rival hypotheses as to the timing of the imprinting of objects of different behavioural routines.

The Conditions of Imprinting

Konrad Lorenz

Further reading

Lorenz, K., *King Solomon's Ring*, transl. M. K. Wilson, London, 1952.

Lorenz, K., *Studies in Animal and Human Behaviour*, transl. R. Martin, vol. 1, London, 1970.

Ardrey, R., *The Territorial Imperative*, London, 1967.

Durant, J. R., 'Innate Character in Animals and Man: A Perspective on the Origins of Ethology', in Webster, C. (ed.), *Biology, Medicine and Society, 1840–1940*, Cambridge, 1981.

Nisbett, A., *Konrad Lorenz*, London, 1976.

Tiger, L., and Fox, R., *The Imperial Animal*, London, 1972.

Tinbergen, N., 'Ethology' in R. Harré (ed.), *Scientific Thought, 1900–1960*, Oxford, 1969, ch. 12.

C

Finding the Form of a Law
Inductively

The laws of nature are not merely qualitative correlations, but, it has turned out, sometimes take very precise forms. These forms are expressed in mathematical relationships revealed by the study of the quantitative aspects of processes – how much, for how long, and so on. Two famous experiments illustrate in a very simple way the kind of work that, through measurement, reveals form. By measuring the times taken for a ball to roll for different distances down a grooved beam **Galileo** was able to formulate precisely one of the laws of accelerated motion, that ratios of distances traversed are directly proportional to the ratios of squares of elapsed time. **Robert Boyle** did not set out so immediately to determine the form of a law. He was generally interested in studying the 'springiness' of gases, and amongst other things in finding out the quantitative relations between the pressures imposed upon and volumes occupied by confined gases. From these results he found a quantitative law.

6. Galileo

The Law of Descent

Fig. 34: Galileo Galilei, chalk drawing by Ottavio Leoni, 1624. Louvre, Paris, Cabinet des Dessins.

Galileo Galilei was born at Pisa on 15 February 1564, the son of Vincenzo Galilei, a cloth merchant. But Vincenzo was also a mathematician and theorist of music, well known in his time. Kepler took Vincenzo's book on harmony with him to read on the journey from Vienna to Graz. Galileo Galilei was partly educated by his father, partly in the monastery at Vallombrosa, near Florence. Advancement in those days depended as much on patronage as it did on talent. Galileo was lucky enough to attract the attention of Marchese Guido Ubaldo del Monte, and was appointed to the chair of mathematics at Pisa, with the help of his patron, when he was still only 25. There is no doubt that Galileo was a tactless and aggressive fellow, and he made many influential enemies. It seems he was rather anxious to leave the poverty and disagreeable conditions of Pisa, and in 1692, through the offices of the same patron, he was appointed to the chair of mathematics at Padua.

Galileo came to prominence in 1610 with the publication of *The Starry Messenger*, an account of a series of remarkable observations made with a telescope of his own development. It included a fairly detailed description of the mountainous terrain of the moon, and above all, a convincing account of his discovery of the moons of Jupiter. It was these very moons, implying a second centre of rotation in the solar system, that began much of Galileo's troubles. They were the objects the Paduan Aristotelians refused to view through his telescope.

In 1610 Galileo came to Florence as chief mathematician to the Grand Duke of Tuscany. Immediately he began to attract a great deal of attention, and acquired friends and admirers in the highest offices of state and Church. In particular he was supported by Pope Urban VIII, whom Galileo had known earlier as Cardinal Bonafeo Barberini. But in 1632, seemingly against the wishes of the pope, he published his *Dialogue on the Two Great World Systems*. In this work the Copernican theory and its rivals are discussed by a group of savants, thinly disguised representations of Galileo and one or two of his acquaintances. Somehow, and just how still remains something of a mystery, Urban VIII was deeply offended by the publication of the book, and arraigned Galileo to appear for trial in Rome. In 1633 Galileo abjured the opinions expressed in the book. He was condemned to house arrest, and forbidden to publish any further works of science. But during his confinement he worked with zeal and vigour on the *Dialogue concerning Two New Sciences*, from which the discoveries described in this section are taken. Of course the book could not be brought out in Italy, but was published by Elzevir, in Leyden, in 1638.

Though he had been somewhat unfeeling about his children in earlier life, in his last years he became very close to his daughter, who cared for him in his failing old age. He died on 8 January 1642.

Early work on the laws of motion: the Merton theorem

The experiment of Galileo, for all its apparent simplicity, was the culmination of work on the laws of motion that had begun in Merton College, Oxford, in 1328. In that year Thomas Bradwardine completed his *Tractatus de Proportionibus*. Bradwardine's interest in the problems of kinematics seems to have stimulated three gifted Mertonian mathematicians, William Heytesbury (*c.* 1310–1380), Richard Swineshead (at Merton in the 1340s) and John Dumbleton (at Merton *c.* 1330 to 1350).

In his history of the science of mechanics in the Middle Ages, Marshall Clagett shows how many basic concepts and theorems of the science of motion were developed by these workers in their mathematical studies. These included the difference between dynamics, the theory of the causes of motion, and kinematics, the theory of the process and effects of motion; the correct formulation of a concept of acceleration, and above all, a proof of the mean-speed theorem, the key to understanding the kinematics of uniformly accelerated motion.

Two central ideas were required. When something is accelerating, it has a different velocity at each instant. This requires the idea of instantaneous velocity, clearly defined by Heytesbury. But if we compare the total distance a moving thing covers with the total time it takes, we can calculate an average or mean velocity. The measurement of instantaneous velocity is impossible since it involves the distance a body *would* cover if it were moved for a standard time at that momentary speed. The stroke of genius that enabled the Mertonian mathematicians to solve the problem of finding the laws of uniformly accelerated motion, was to show that the effects of accelerated motions could be worked out in terms of average or mean speeds.

What then was the 'mean-speed theorem'? A uniformly accelerating body will cover a distance equal to what it would have covered in the time, if it had been moving uniformly at its mean or average velocity. Simplifying the picture by supposing that a body starts from rest, the theorem can be expressed geometrically.

Scholars differ on how far Galileo took his hypothesis of the form of the law of uniformly accelerated motion directly or indirectly from these mathematical analyses. In the *Two New Sciences* Galileo is quite explicit. He says (Drake translation, p. 169) that he did the experiments 'in order to be assured that the acceleration of heavy bodies falling naturally does follow the ratio expounded above ...' And that exposition is a proof of the mean-speed theorem. However, Stillman Drake, on the basis of a study of Galileo's working notes, has suggested that in 1603 or 1604 Galileo carried out an experiment with a ball rolling down an inclined plane, and that 'he had no inkling of the law before he made the experiment' (Drake, 1978, pp. 84–90; see Further Reading). Whatever may be the truth of the matter the experiment I am about to describe takes for granted that there is a law whose precise form must be found.

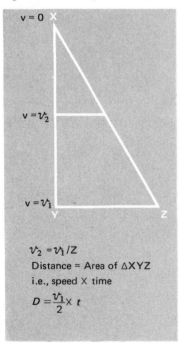

Fig. 35: The mean speed theorem.

$v_2 = v_1 / 2$

Distance = Area of $\triangle$XYZ

i.e., speed $\times$ time

$$D = \frac{v_1}{2} \times t$$

Galileo's experimental discovery of the form of a kinematic law

Galileo carefully distinguishes between the mathematical study of motion and the empirical study of movement. 'Anyone', he says, 'may invent an arbitrary type of motion and discuss its properties. We have decided to consider the phenomena of bodies falling with an acceleration such as actually occurs in nature ... in the belief [that we have done so] we are confirmed mainly by the consideration that experimental results are seen to agree with ... those properties which have been demonstrated by us.' The first thing to notice is that heavy bodies start falling slowly and gradually increase their speed, in short, they accelerate. This can easily be demonstrated by dropping a heavy ball on to a cushion from a greater and greater height. The longer it is falling, the deeper the dent made in the cushion. But in free fall the motion of bodies is very difficult to observe and measure precisely. The trick is to transfer the motion to an inclined plane and so to investigate motion under a more gradual acceleration than that of gravity. The mean-speed theorem implies that the ratios of the distances travelled is proportional to the square of the times taken for those distances. Whether he had indeed derived his ideas of the law from that theorem or from prior experiment, Galileo set about comparing the ratios of distances travelled with the ratios of times taken.

170 DIALOGO TERZO

tibus temporis A B maximus & ultimus repræsentetur per E B, utcunque super A B constituta : junctæque A E lineæ, omnes ex singulis punctis lineæ A B ipsi B E æquidistanter actæ crescentes velocitatis gradus post instans A repræsentabunt. Divisa deinde B E bifariam in F, ductisque parallelis F G, A G, ipsis B A, B F ; Parallelogrammum A G F B erit constitutum triangulo A E B æquale, dividens suo latere G F, bifariam A E in I : quodsi parallelæ trianguli A E B usque ad I G F extendantur, habebimus aggregatum parallelarum omnium in quadrilatero contentarum æqualem aggregatui comprehensarum in triangulo A E B. quæ enim sunt in triangulo I E F, paria sunt cum contentis in triangulo G I A ; ex vero quæ habentur in trapezio A I F B, communes sunt. Cumque singulis & omnibus instantibus temporis A B respondeant singula & omnia puncta lineæ A B, ex quibus actæ parallelæ in triangulo A E B comprehensæ crescentes gradus velocitatis adauctæ repræsentant ; parallelæ vero intra parallelogrammum contentæ totidem gradus velocitatis non adauctæ, sed æquabilis, itidem repræsentent : apparet totidem velocitatis momenta absumpta esse in motu accelerato juxta crescentes parallelas trianguli A E B, ac in motu æquabili juxta parallelas parallelogrammi G B : quod enim momentorum deficit in prima motus accelerati medietate, (deficiunt enim momenta per parallelas trianguli A G I repræsentata,) reficitur à momentis per parallelas trianguli I E F repræsentatis. Patet igitur, æqualia futura esse
spatia

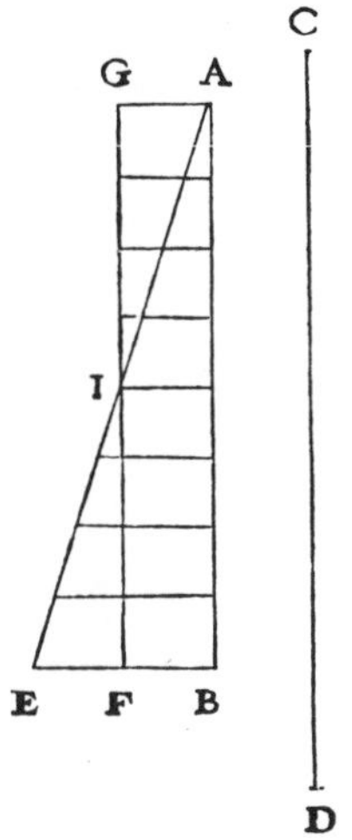

Fig. 36: Geometrical illustration of mean speed theorem from Galileo's *Dialogues concerning Two New Sciences*, Leyden (1638), p. 170.

The experiment involved cutting and polishing a groove in a wooden beam and lining the groove with parchment. A polished bronze ball was let roll down the groove when the beam was set on an incline. In the first range of experiments the amount of variation to be expected in such a series of trials was tested by measuring the time of whole descents, using the pulse as timing device. Variations in time for many runs of the same descent were very small.

The theoretically derived relation between distances and

Fig. 37: Galileo demonstrating his apparatus. Contemporary fresco by Giuseppe Bezzuoli. Tribuna di Galileo, Museo Zoologico 'La Specola', Florence.

Fig. 38: A replica of Galileo's apparatus, made in 1775 for the Grand Duke of Tuscany. Istituto e Museo di Storia della Scienza, Florence.

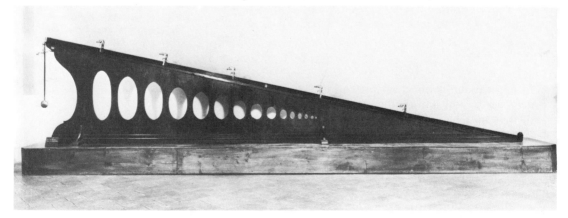

times for uniformly accelerating motion was tested by letting the ball roll a quarter, then half, then two-thirds and so on, of the length of the groove, measuring the times for the journey in each case. The ball did indeed take half the time required for a full descent to reach the quarter way point. And whatever the distance chosen, 'repeated a full hundred times, we always found that the spaces traversed were to each other as the squares of the times.' In the final series of experiments time was measured by the weight of water that escapes through a thin tube fixed in the bottom of a vessel so large that the loss of water did not sensibly affect the pressure in the escape tube, and so did not alter the rate at which water escaped.

T. Settle (1961) has repeated this experiment in a manner as similar to Galileo's original method as possible. He was not only able to replicate Galileo's own results rather well, but in so doing he put paid to the once prevalent view that Galileo's experiments were mostly imaginary.

Subsequent developments in the science of motion

But two questions remain unanswered by Galileo's investigations. Why do bodies fall with uniform acceleration? Can the terrestrial laws of motion be applied to all the bodies in the universe, including the stars and the planets? One set of answers was supplied by Newton, that satisfied the scientific community until the beginning of the twentieth century.

Following Kepler, Newton supposed that there were forces acting between the centres of any two material bodies in the universe. These were the effect of an unexplained influence, gravity. Newton proposed a fundamental principle, the law of gravity. The gravitational force acting between any two bodies is inversely proportional to the square of the distance separating them, and directly proportional to the product of their masses. Over small distances such as those through which bodies fall on the surface of the earth, this force is relatively constant and produces a uniform acceleration, the increasing speed of fall that Galileo had studied.

The gravitational law explained why the moon orbited the earth and the planets the sun. These bodies would have a tendency to fly off in straight lines at a tangent to their orbits if there had been no gravity. But because they are subject to gravitational force, they are drawn towards the heavy body around which they turn. In short they are forever falling. It is the combination of the tendency to fly off with a tendency continuously to fall, that is exactly balanced in an orbiting body. This accounts for the very many cases of near-circular orbital motion that we find in the heavens. The same laws apply everywhere, among the stars as on earth.

In the centuries that followed Galileo's demonstration that

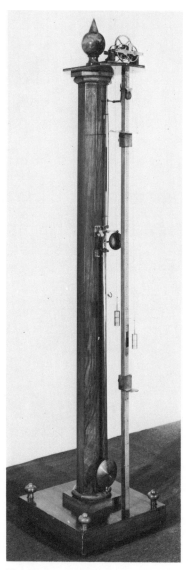

Fig. 39: Atwood's fall machine
(c.1830). Whipple Museum of the
History of Science, Cambridge.

the mathematical analysis of motion begun by the Mertonians was applicable to the real world, there was a fairly steady progressive refinement of concepts and elaboration of more sophisticated mathematical methods. Energy and momentum were distinguished, and the calculus replaced geometry as the main tool of analysis. These developments allowed for more complex motions and more elaborately structured mechanisms to be mathematically represented.

More sophisticated machines for testing the applicability of the laws of mechanics to nature were developed in the nineteenth century, notably Atwood's machine.

In Galileo's experiment we have a very pure case of the demonstration of the applicability of a conceptual system to the real world, a system which was developed in thought. The rationale of the experiment could be given in neither the inductivist nor the fallibilist theory.

Further reading

Galilei, G., *Dialogues Concerning Two New Sciences*, transl. Stillman Drake, Madison, Wisc., 1974 (original publication, Elzevir, Leyden, 1638).

Clagett, M., *The Science of Mechanics in the Middle Ages*, Madison, Wisc., 1961.

Drake. S., *Galileo at Work*, Chicago and London, 1978.

McMullin, E. (ed.), *Galileo Man of Science*, New York, 1967.

Santillana, G., *The Crime of Galileo*, Chicago, 1955.

Settle, T., 'An Experiment in the History of Science', *Science*, 133, 1961, pp. 19–23; and see also MacLachlan, J., *Scientific American*, March 1975, pp. 109–10.

7. Robert Boyle

The Measurement of the Spring of the Air

Robert Boyle was born in Lismore, in Ireland, in 1627. Though the youngest son of a family of fourteen, he grew up in considerable affluence. His father was the first Earl of Cork. Robert Boyle's mother was the Earl's second wife. At the age of eight he was sent off to boarding school, to Eton, just then beginning to be fashionable for the education of the sons of gentlefolk. He was at Eton for four years, and subsequently in Geneva, where he devoted a great deal of attention to mathematics.

It was there that he decided to devote himself to science. One evening he was watching a spectacular display of lightning, and began to wonder why he was not struck. He came to the conclusion that God must have reserved him for some special task. With the emphasis on natural religion in that time, it was not surprising that he dedicated himself to the demonstration of God's majesty by unravelling the secrets of nature. From Geneva Boyle travelled to Italy, and spent some time in Florence. There he studied the works of Galileo.

The outbreak of the Civil War led him to return to England. He might have been expected to have Royalist leanings, but for a variety of reasons he had Parliamentarian sympathies. These brought him into contact with Samuel Hartlibb. Through this friendship Boyle was encouraged to study medicine. It was during his efforts to prepare drugs and medicines that he began to take an interest in chemistry.

In 1656 he settled in Oxford, in a house next to University College, on a site that now boasts the grotesque Shelley Memorial. Here he worked to provide experimental proofs of the corpuscularian, mechanical theory of nature. He became friendly with the leading mathematicians of the time, Wallis and Ward. Perhaps more importantly, he joined the circle around John Locke at Christ Church, in discussions of the philosophical basis of the mechanical theory of nature. This

Fig. 40: A contemporary engraving of Robert Boyle. Museum of the History of Science, Oxford University.

Robert Boyle

Left Fig. 41: Instruments for the study of the spring of the air, from Boyle's *Continuation of New Experiments Physico-Mechanical . . .* , Oxford (1669). The calibrated tube within the jar measures the expansion of the air in the other arm when the pressure in the jar is reduced. The rest of the apparatus measures the drop in pressure as air is pumped out.

Right Fig. 42: More instruments for the study of the spring of the air, illustrated in *The Works of the Hon. Robert Boyle*, ed. T. Birch, London (1772), vol. I, plate i. Fig. 1 is an air pump. Fig. 16, actually used in an experiment to determine the relative weight of water and quicksilver, is very similar to the classic Boyle's Law apparatus, except that the end of the shorter section of the tube should be closed.

phase of his scientific activities was summed up in his famous work, *The Origine of Formes and Qualities*, published in 1666.

After the Restoration he moved to London, taking a very active part in the founding of the Royal Society. His intensely religious attitude to the world involved him in a number of projects for the propagation of religion. Throughout his career he had written small, entertaining tracts, and even ventured one of the first historical novels in English, *The Martyrdom of Theodora*, on the theme of the conflict between personal love and religious duty. He died in London in 1691.

The study of gases prior to Boyle

The problem motivating most studies of 'airs' in the seventeenth century concerned the nature and even the possibility of the vacuum. Orthodox opinion denied that a really empty space was physically possible since 'nature abhorred a vacuum'. By filling a long tube with mercury and inverting it over a dish of the same liquid, Torricelli had shown that at the upper end of a closed tube a vacuum is formed as the mercury drops to a level which the weight of the air will support. Why is this 'factitious' or manufactured vacuum not found in nature? Those who believed that vacua were possible, and particularly

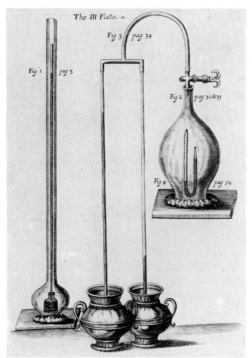

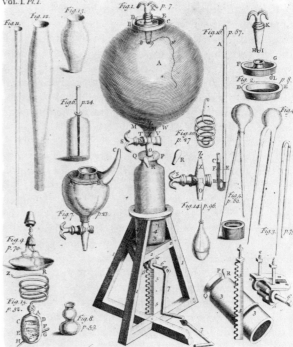

that Torricelli had demonstrated their actual existence, had to explain why there was a tendency to fill all empty spaces so that vacua were rare and unstable. Boyle was among those who believed this was due to a real expansive power of the air.

The beginnings of an experimental investigation of the problem had been made by von Guericke. He made two hemispheres of brass, which fitted nicely together. Each was harnessed to a team of horses. The air was expelled from inside the pair of half globes by the steam from boiling water. When this condensed, a vacuum formed within the hemispheres. Air pressure on the outsides kept the spheres together so well that even two teams of horses could not separate them. Still, the reality of the expansive power of the air had not been directly verified.

Boyle's first set of experiments were designed to demonstrate the active power of the air directly. In *New Experiments, Physico-mechanicall, touching the Spring of the Air* (3rd edn., p. 2), Boyle says, 'Divers ways have been proposed to show both the Pressure of the Air, as the Atmosphere is a heavy Body, and that Air, especially when compressed by outward force, has a Spring that enables it to sustain or resist equal to that of as much of the atmosphere, as can come to bear against it, *and* also to show, that such Air as we live in, and is not condensed by any human or adventitious force, has not only a *resisting* Spring, but an active Spring (if I may so speak) in some measure, as when it distends a flaccid or breaks a full-blown bladder in our exhausted Receiver.'

But a more direct experiment was wanted. To demonstrate the active spring of the air as a phenomenon Boyle and Hooke set up apparatus similar to that to be seen in Figure 41. Inside the large tube from which the air can be extracted, is a smaller tube with mercury trapped within it. This mercury compresses some trapped air. While the outer tube is full of air the pressures balance one another. But when the outer tube is evacuated the trapped air actively thrusts out the mercury from the tube above it. I suppose that a quite spectacular fountain of mercury sprayed up out of the inner tube, as the air was suddenly extracted.

To complete the study of the air as a spring Boyle proposed to make a 'measure of the Force of the Spring of the Air compressed and dilated', that is to measure accurately how the spring increased when the column of air was decreased by pressure, and how it decreased when the volume was increased by lowering the outside pressure.

The experiment: measuring the spring of the air

The apparatus was relatively simple. Boyle and his assistant, Hooke, took a long glass tube 'crooked at the bottom' with 'the

Fig. 43: An air pump made *c.*1740. It is of a type similar to those produced by Denys Papin (1647–1712), who worked with Boyle. Two pistons are operated by a rack-and-pinion mechanism. The glass bell-jar to be seen in fig. 42 on p. 84 is missing. Museum of the History of Science, Oxford University.

Robert Boyle

orifice of the shorter leg ... being hermetically sealed'. They carefully pasted strips of paper along each leg, and marked them in inches. The tube was filled with mercury from the open longer end. Air was allowed to pass out from the closed end by 'frequently inclining the tube' so that 'the air in the enclosed tube should be of the same laxity as the rest of the air about it'. Then with the pressures equalized they began to pour mercury in the open end to increase the pressure on the enclosed air. They continued until the enclosed air was reduced to half its original volume.

By using what Boyle calls the Torricellian tube, which we would call a barometer, he and Hooke had measured the air pressure obtaining during the experiment, the equivalent of 29 inches of mercury. When the volume of enclosed air had been reduced to one half, the additional 'head' of mercury in the open end of the tube measured just 29 inches. In short 'this observation does both very well agree with and confirm our hypothesis ... that the greater the weight is, that leans upon the air, the more forcible is its endeavour of dilation and consequently its power of resistance (as other springs are stronger when bent by greater weights).' At this point the tube broke. They tried again with a new tube of a 'pretty bigness'.

Using the newer, stronger tube they were able to make a series of observations examining the relation between the 'endeavour' of the air measured by the weight of mercury required to compress it, and the volume to which the original air had been reduced. The results are shown in the table.

reducing!

Fig. 44: The results of compressing the air, from Boyle's *Defence* of his *New Experiments* against the objections of Franciscus Linus. It was in this work, appended to the 2nd edition of the *New Experiments* in 1662, that Boyle first published the tables showing his 'Law' of the reciprocal relation between the pressure and volume of a gas. *Works*, ed. Birch, vol. I, p. 260.

A table of the rarefaction of the air.

A. The number of equal spaces at the top of the tube, that contained the same parcel of air.

B. The height of the mercurial cylinder, that together with the spring of the included, air counterbalanced the pressure of the atmosphere.

C. The pressure of the atmosphere.

D. The complement of B to C, exhibiting the pressure sustained by the included air.

E. What that pressure should be, according to the hypothesis.

A	B	C (Subtracted from $29\frac{1}{4}$ leaves)	D	E
1	$00\frac{0}{0}$		$29\frac{3}{4}$	$29\frac{1}{4}$
$1\frac{1}{2}$	$10\frac{5}{0}$		$19\frac{1}{8}$	$19\frac{5}{0}$
2	$15\frac{3}{8}$		$14\frac{3}{8}$	$14\frac{7}{8}$
3	$20\frac{2}{8}$		$9\frac{4}{8}$	$9\frac{15}{32}$
4	$22\frac{2}{8}$		$7\frac{1}{4}$	$7\frac{7}{16}$
5	$24\frac{5}{8}$		$5\frac{5}{8}$	$5\frac{19}{20}$
6	$24\frac{7}{8}$		$4\frac{7}{8}$	$4\frac{27}{6}$
7	$25\frac{2}{3}$		$4\frac{2}{8}$	$4\frac{1}{4}$
8	$26\frac{0}{0}$		$3\frac{6}{8}$	$3\frac{21}{32}$
9	$26\frac{1}{8}$		$3\frac{1}{8}$	$3\frac{11}{36}$
10	$26\frac{5}{8}$		$3\frac{0}{0}$	$2\frac{39}{40}$
12	$27\frac{1}{8}$		$2\frac{5}{8}$	$2\frac{13}{48}$
14	$27\frac{3}{8}$		$2\frac{2}{8}$	$2\frac{1}{9}$
16	$27\frac{6}{8}$		$2\frac{0}{0}$	$1\frac{55}{64}$
18	$27\frac{7}{8}$		$1\frac{7}{8}$	$1\frac{47}{73}$
20	$28\frac{0}{0}$		$1\frac{6}{8}$	$1\frac{9}{10}$
24	$28\frac{2}{8}$		$1\frac{4}{8}$	$1\frac{23}{30}$
28	$28\frac{3}{8}$		$1\frac{3}{8}$	$1\frac{1}{16}$
32	$28\frac{4}{8}$		$1\frac{2}{8}$	$0\frac{119}{173}$

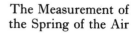

Fig. 45: Artist's impression of Boyle's experiment, with precautions against the tube breaking.

It is worth noticing that the experiment is not designed to discover what happens to air when it is subjected to a compressing force, but to find how the force exerted by the air is related to its state of compression. It is an attempt to measure the active power of air to resist force, its spring.

The experiment was hedged around with precautions. Boyle and Hooke placed the bottom end of the tube in a wooden box, not only to catch 'sipplings' of mercury, but in case the tube broke again. From the way Boyle puts this I suspect that they had not taken this precaution with the first experiment, and when the tube broke found the quicksilver all over the floor.

The experiment had been done at room temperature. What would be the effect of heating or cooling the trapped air? By putting a wet cloth around the tube they hoped to cool it, but 'it sometimes seemed a little to shrink, but not so manifestly that we dare build anything upon it'. However, when they cautiously heated the closed end with a candle flame, 'the heat had a more sensible operation'. The table involves figures that do not exactly conform to a law of strict proportionality. But errors 'may probably enough be ascribed to such want of exactness as in such nice experiments is scarce avoidable'.

Boyle was very well aware of the problem of formulating universal hypotheses on the basis of a few experiments. 'But, for all that,' he says, 'I shall not venture to determine whether or no the intimated theory will hold universally and precisely . . .' 'No one perhaps yet knows how near to an infinite compression the air may be capable of, if the compressing force be competently increased.' It was to just this question that Amagat, as we shall see, eventually provided an answer.

But the experiment as described tested only the effect of increasing the pressure on the air to greater than that produced by the weight of the atmosphere. There should be a corresponding reduction in pressure for air which has expanded beyond its normal volume. The apparatus had to be different, since they had no flexible tubes by which the surface of the mercury could be lowered. 'We provided', says Boyle, 'a slender glass-pipe of about the bigness of a swan's quill.' They glued a paper strip with inches marked along it to the tube. The little tube was inserted into a wide tube, filled with mercury so that about one inch protruded above the surface. The pipe was sealed with wax to trap an inch of air within. 'After which the pipe was let alone for a while, that the air, dilated a little by the heat of the wax, might, upon refrigeration, be reduced to its wonted density.' By lifting up the slender pipe the air within was subjected to decreasing pressure, so that it was dilated to 1½ inches, 2 inches and so on.

compressing!

Fig. 46: The results of reducing the pressure of the air. Table from Boyle's *Defence, Works*, ed. Birch, vol. I, p. 158.

A table of the condenſation of the air.

A	A	B	C	D	E
48	12	00	Added to $22\frac{1}{8}$ makes	$29\frac{2}{16}$	$29\frac{2}{16}$
46	$11\frac{1}{2}$	$01\frac{7}{16}$		$30\frac{9}{16}$	$33\frac{6}{16}$
44	11	$02\frac{13}{16}$		$31\frac{15}{16}$	$31\frac{12}{16}$
42	$10\frac{1}{2}$	$04\frac{6}{16}$		$33\frac{8}{16}$	$33\frac{1}{7}$
40	10	$06\frac{3}{16}$		$35\frac{5}{16}$	35- -
38	$9\frac{1}{2}$	$07\frac{14}{16}$		37	$36\frac{15}{19}$
36	9	$10\frac{2}{16}$		$39\frac{5}{16}$	$38\frac{7}{8}$
34	$8\frac{1}{2}$	$12\frac{8}{16}$		$41\frac{10}{16}$	$41\frac{2}{17}$
32	8	$15\frac{1}{16}$		$44\frac{3}{16}$	$43\frac{11}{16}$
30	$7\frac{1}{2}$	$17\frac{15}{16}$		$47\frac{1}{16}$	$46\frac{3}{5}$
28	7	$21\frac{3}{16}$		$50\frac{5}{16}$	50- -
26	$6\frac{1}{2}$	$25\frac{3}{16}$		$54\frac{5}{16}$	$53\frac{10}{13}$
24	6	$29\frac{11}{16}$		$58\frac{13}{16}$	$58\frac{2}{8}$
23	$5\frac{3}{4}$	$32\frac{3}{16}$		$61\frac{5}{16}$	$60\frac{18}{23}$
22	$5\frac{1}{2}$	$34\frac{15}{16}$		$64\frac{1}{16}$	$63\frac{6}{11}$
21	$5\frac{1}{4}$	$37\frac{15}{16}$		$67\frac{1}{16}$	$66\frac{4}{7}$
20	5	$41\frac{9}{16}$		$70\frac{11}{16}$	70- -
19	$4\frac{3}{4}$	45- -		$74\frac{2}{16}$	$73\frac{11}{19}$
18	$4\frac{1}{2}$	$48\frac{12}{16}$		$77\frac{14}{16}$	$77\frac{2}{3}$
17	$4\frac{1}{4}$	$53\frac{11}{16}$		$82\frac{12}{16}$	$82\frac{4}{17}$
16	4	$58\frac{2}{16}$		$87\frac{14}{16}$	$87\frac{3}{8}$
15	$3\frac{3}{4}$	$63\frac{15}{16}$		$93\frac{1}{16}$	$93\frac{1}{5}$
14	$3\frac{1}{2}$	$71\frac{5}{16}$		$100\frac{7}{16}$	$99\frac{6}{7}$
13	$3\frac{1}{4}$	$78\frac{11}{16}$		$107\frac{13}{16}$	$107\frac{7}{13}$
12	3	$88\frac{7}{16}$		$117\frac{9}{16}$	$116\frac{4}{8}$

AA. The number of equal ſpaces in the ſhorter leg, that contained the ſame parcel of air diverſly extended.

B. The height of the mercurial cylinder in the longer leg, that compreſſed the air into thoſe dimenſions.

C. The height of the mercurial cylinder, that counterbalanced the preſſure of the atmoſphere.

D. The aggregate of the two laſt columns *B* and *C*, exhibiting the preſſure ſuſtained by the included air.

E. What that preſſure ſhould be according to the hypotheſis, that ſuppoſes the preſſures and expanſions to be in reciprocal proportion.

They had already found that the barometric pressure was 29¾ inches that day, and to their satisfaction the difference between the levels of the tube when the air was dilated to double its original volume was only half the height of the barometer.

An error was found. When they replunged 'the pipe into the quicksilver' the air had slightly gained in volume at atmospheric pressure. Boyle supposed that this increase had come from 'little aerial bubbles in the quicksilver, contained in the pipe (so easy is it in a nice experiment to miss of exactness)'.

Studies of gases after Boyle

The development of gas experiments followed three distinct lines. Boyle and Hooke had studied only air, and that at low pressures and low temperatures. When Andrews subjected carbon dioxide to moderate pressures he found that below a certain temperature the gas no longer obeyed Boyle's Law.

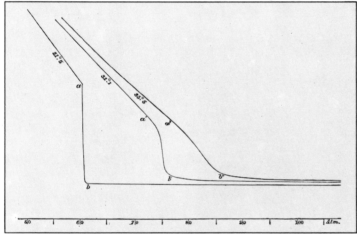

Fig. 47: Relations between pressure and volume showing deviations from Boyle's Law. Near the 'critical point' where the curve is parallel to the pressure axis, the gas has liquefied. T. Andrews's Bakerian Lecture on 'The Gaseous State of Matter', *Philosophical Transactions of the Royal Society*, vol. 166 (1876), p. 443.

Indeed below the 'critical temperature' the gas liquified as the pressure increased without any further cooling. These studies were vastly expanded by E. H. Amagat. He had begun by lowering a long tube down his father's coalmine. By this means he obtained very great pressures. Later he developed mechanical methods of compression to as high as 400 times the atmospheric pressure, in apparatus that allowed him to vary temperature systematically. He found that with the exception of hydrogen all gases exhibited, in some degree, the deviations found by Andrews. Boyle's Law had its limits.

Amagat was no positivist, satisfied with a mere correlation – he tried to explain why gases under high pressure did not obey Boyle's Law. By borrowing some ideas of Clausius to the effect that gases should be thought of as swarms of randomly moving

Robert Boyle

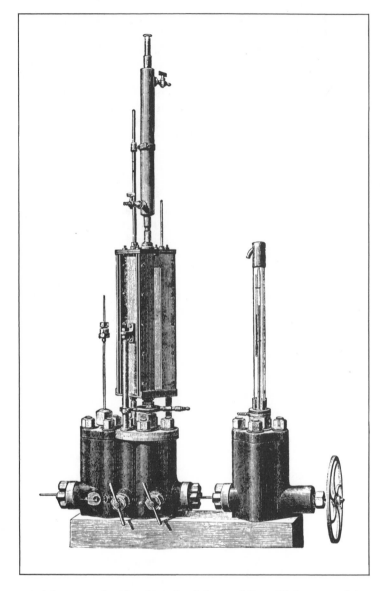

Fig. 48: Amagat's apparatus.
Illustration from E. H. Amagat, *The
Laws of Gases* (1899), p. 17.

particles or molecules, he solved the problem. If these particles
were real they should have a volume, say a. Then the true
available volume in which gas molecules can move is not V, the
volume of the container, but $V - a$, the container volume
reduced by the volume taken up by the molecules. Even if the
gas is subject to infinite pressure it cannot be compressed to
less than the molecular volume. A simple mathematical
relation can express this:

$$p \ (V - a) = \text{constant}.$$

If we divide through by p we get

$V - a = \text{constant}/p$.

When $V = a$ under complete compression we get

$0 = \text{constant}/p$,

and since any number divided by infinity is 0, p must be infinite. Boyle's question about the universality of his law had been solved. When Amagat analysed his experimental results he found that they more nearly followed the shape of the curve obtained by plotting the values of p and V in the equation $p(V - a) = \text{constant}$ than any other reasonable equation.

Although Robert Norman and Galileo had both made measurements as important steps in their work, even in the case of Galileo's rolling ball measurement was not quite the central point. Galileo knew what the law of descent had to be. The measurements merely clinched the truism that what *must* be surely *is*. In Boyle's experiment the law emerged from the measurements. It seems fairly likely that Boyle knew roughly what sort of law to expect, but throughout the history of the scientific study of the properties of gases, the measurements have had the final say. Amagat did not show that Boyle's Law was wrong – he showed that it followed from his more general theory as a special case. If the volume of gas is so great that we can ignore the volume of the molecules themselves then the new law reduces to the old.

Further reading

Boyle, R., *New Experiments, Physico-mechanicall, touching the Spring of the Air and its Effects*, London, 1660; 2nd edn., incorporating Boyle's *Defence*, 1662; 3rd edn., 1682.

Bacus, C. (ed.), *Memoirs on the Laws of Gases*, New York, 1899.

Hall, M. B. (ed.), *Robert Boyle on Natural Philosophy*, Bloomington, Ind., 1965.

Maddison, R. W., *The Life of the Honourable Robert Boyle*, London, 1969.

D

The Use of Models to
Simulate an otherwise
Unresearchable Process

In the examples described so far it was possible for the experimenter to work directly on the natural process under study – the flow of sap, accelerating bodies, developing embryos and so forth. But there are processes that are remote from observation or experimental manipulation. Yet they may have a key role in the production of a puzzling natural effect. To deal with cases like this scientists create physical models of the systems involved in the process they are studying; by manipulating the model and seeing how it behaves they infer corresponding processes in the real thing. One of the earliest and most satisfying uses of models in experiments was made by **Theodoric of Freibourg**, when he used glass globes to simulate the role of raindrops in the formation of the rainbow.

8. Theodoric of Freibourg

The Causes of the Rainbow

Theodoric was born somewhere in Germany, probably a little before 1250. It is known that he studied in Paris from 1275 to 1277. He was a member of the Order of Preachers, the Dominicans. He seems to have had a very successful career in his Order, holding the high office of Provincial of Germany from 1293 to 1296. He was present at the General Chapter held at Toulouse in 1304. It was there that Aymeric, at that time Master General of the Dominicans, suggested to Theodoric that he make a systematic study of the rainbow. This fact helps us to date his major work, the *De Iride* (*On the Rainbow*), in which he wrote up the results of his studies of light. It must have been composed during the time that Aymeric was Master of the Order, that is between 1304 and 1311. According to Theodoric's own account he gave up teaching in later life to devote himself to Church ministry. It seems likely that he had completed his scientific work before this change of vocation. He was present at the General Chapter of the Order at Piacenza in 1310. He probably died shortly afterwards.

William Wallace, the best modern biographer of Theodoric, describes him as a man of a somewhat independent turn of mind. Religious orders in those days were strictly disciplined and not inclined to encourage individuals to pursue private interests. Wallace suggests that this may account for Theodoric's apparent reticence in publishing his researches. That this independent standpoint was not confined to science is evidenced by the fact that he was widely credited with being the first scholastic to preach in the vernacular, German. The scientific investigations reported in *De Iride* are outstanding for the degree to which Theodoric subjects every point, whether derived from ancient sources or from an idea of his own, to scrupulous empirical test. The work does not suggest a passive, merely scholastic acceptance of traditional authorities.

Though much of medieval science was a mere repetition of

Fig. 49: A medieval scientist at work: Richard of Wallingford, an English contemporary of Theodoric, depicted marking out a metal plate for an astronomical instrument. British Library, MS Cotton Claudius E. iv. No portrait of Theodoric is known.

material derived in large part from the works of Aristotle, a good deal of work of the highest quality was undertaken, here and there. In the domain of experimental science Theodoric's study of the rainbow is, to my mind, the most impressive to come down to us from that time. Furthermore, in its basic essentials, it remains the accepted account of the formation of the rainbow.

The state of rainbow studies before Theodoric

The problem of explaining the rainbow focuses attention on two important issues in the understanding of light and its effects. How are the colours formed? What is the explanation of the striking geometrical regularities to be seen in the phenomena of reflection and refraction? The case of the rainbow offers these problems in very particular form. Why are the colours formed in the order in which they are always found? Why does the rainbow have such a very specific and unvarying geometrical form? Why is it always an arc of a circle and why is the highest point of the arc always at the same angle of elevation above the horizon? In these questions the problems facing any student of light are summed up, how to account for colour and how to explain the geometry of light.

In the *Meteorologica* (Book 3) Aristotle had proposed that the appearance of the rainbow is due to reflection from newly formed raindrops which form a 'better mirror than mist'. Some medieval commentators had proposed that the circular form of the bow is simply a reflection of the circular disc of the sun. Most assumed that the phenomenon is essentially one of reflection, with the falling raindrops acting as a mirror. Albertus Magnus first proposed the theory that the rainbow was produced by light interacting with each drop. This idea brings in the spherical shape of the drop for the first time. But Albertus thought that the colours were produced somehow within the curtain of drops, by the effects of some kind of layering. At about the same time as Theodoric was carrying out his masterly experimental investigation, Peter of Alvernia suggested that the rainbow is due to refraction rather than reflection.

Arguments about the colours had turned on whether they were really there in the sky as coloured bands, or whether they were some kind of subjective effect. Most commentators seem to have thought that the colours were real, produced in an interaction between light from the sun and the falling drops.

The rainbow resolved: the experiment with water-filled urine flasks

Theodoric set out to investigate both aspects of the rainbow, that is the origin of the order and hue of the colours in the bow, and the source of its very particular geometry. Each step was controlled by a theory, and each stage in the development of the theory was rigorously tested by experiment or observation. As we shall see, his theory of colours was wildly wrong in detail, though carefully and honestly 'verified' by experiment. But in one central particular he was right, that is he held, correctly, that colours were formed in interaction with the water drop.

Theodoric's explanation of how colours are generated is very complex, and I shall present a somewhat simplified version of it here. He believed that there were four radiant colours, red, yellow, green and blue, and that they were distinct. So he did not recognize a continuous spectrum as we do today. Those influenced by Greek thought, and particularly by Aristotle, framed their theories in terms of contraries. Four distinct colours could be produced by two pairs of contraries. Theodoric based his theory on the contrary properties of a medium: whether it is bounded or unbounded, and whether it is clear or opaque. Red and yellow are clear colours, green and blue are obscure. Perhaps he could be taken to mean that the former are nearer to bright white and the latter to dull black. To explain how these distinct colours are generated he argues that where light is received in a bounded region of a medium such as glass, the clear colour will be red, and in an unbounded region, yellow. A glass prism is more bounded near the surface and less bounded deeper within, hence the ray that passes closest to the surface will be red, and the deeper one will be yellow. In the case of the obscure colours it is the relative opacity of the medium that is responsible for the production of distinct hues. Where the medium is more opaque, blue will be produced, where it is more transmissive, green.

The next step is to apply this theory of the production of colours to the passage of light through transparent prisms, spheres and so on. Theodoric undertook a well-planned series of experiments to test each aspect of the theory. Any translucent body is more opaque in the interior than near the surface. If light is refracted in such a body, say a glass prism, the clearer colours will be produced nearer the surface, since a medium becomes more opaque in its depths. Hence red and yellow will be produced in that part of the medium that is nearest the surface and blue and green in the deeper parts. Taking the supposed effects of boundedness and unboundedness with the distinction between the more transmissive and more opaque parts of the medium, we get the four colours in the order red, yellow, green and blue.

The experimental verification of the predicted order of colours came with Theodoric's experiments with a hexagonal prism and a large water-filled glass globe. A. C. Crombie has suggested that this might have been a urine flask as used in medicine. The passages of the rays of light through the medium are carefully drawn in Theodoric's diagrams, preserved in a manuscript in Basle University Library. In the first illustration it can be seen clearly how Theodoric came to think that red, the clearest colour, was produced nearer the surface in the more bounded part of the prism, while blue, the most obscure, is produced in its depths, where the medium is most opaque.

But in the production of the colours of the rainbow there is another, intermediary process. If one looks at a rainbow the uppermost colour is red, the lowest blue. The discovery of the cause of this particular order of colours is Theodoric's master experiment. He shows that to get the effect from a spherical drop of water, the light must be both refracted at the surface and reflected on the inside of the drop. To study this phenomenon he used a model of a raindrop, a large water-filled flask, so that he could study the phenomenon in his laboratory, so to speak. The path of the rays can be seen clearly in the second diagram. The order of colours is reversed, because of the internal reflection. We can see that the order of colours in the rainbow is not incompatible with the basic theory of their production. The red is produced nearer the surface, when the ray passes across the drop. Blue, as an obscure colour, is generated deeper in the medium. But in the reflection, there is a geometrical reversal of the rays which have already acquired their colours.

Notice the logic of this experiment. It is a correct demonstration of two important facts that are still part of the corpus of accepted scientific knowledge. Theodoric showed that light rays of determinate colours travelled specific pathways within the drop. The order of the colours was the effect of these differentiated paths. He showed, too, that the hues were produced, somehow, in the drop itself, not in the eye of the beholder. This too is correct, though we place no credence on his theory of how this happens. We no longer think of the boundedness and unboundedness of different parts of media, nor of the distinction between clear and obscure colours, as physically significant.

Sometimes the primary rainbow is accompanied by a secondary bow, in which the order of colours is reversed. By demonstrating the possibility of a second internal reflection within the drop Theodoric was able to explain how that bow was formed and why its colours were reversed.

Left Fig. 50: The passage of rays of light through a prism. Illustration from the Basle manuscript of the *De Iride*, Universitäts-Bibliothek, F. IV. 30, fol. 15.

Right Fig. 51: Rays of light in a spherical drop. *De Iride*, Basle manuscript, fol. 21.

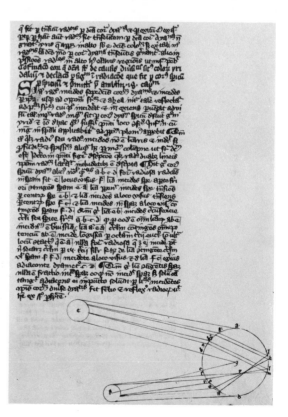

Theodoric of Freibourg

Having shown that the phenomenon of the colours can be explained by refraction and internal reflection, and demonstrated the paths of light within the drops, Theodoric went on to offer a geometrical analysis of the structural properties of the bow. The first step is to argue that the paths of light found within the spherical flask are the same as those within the real rain drops. This is a reasonable supposition if we accept Albertus's suggestion that the drops are falling so fast that they can be thought to be replacing each so rapidly that they are equivalent to a curtain of stationary transparent globes. The general geometry of the rainbow now follows simply by applying the construction for individual drops.

Unfortunately Theodoric's construction is based upon a serious error. In his diagram the sun is represented as if it were distant from the observer by roughly the same order of magnitude as the observer is from the raindrops. The circle on which the drops are represented includes the sun. For a correct construction the sun must be taken to be infinitely distant and the rays as parallel to one another.

But true to his predilection for experimental verification Theodoric measured the angle of greatest elevation of the bow. In several places he says that the measured value is 22°. This is the angle θ in the explanatory diagram. This is a curious error, since it was well known that the correct value for the elevation of the summit of the bow is about double that figure. In his

Fig. 52: Illustration of Theodoric's explanation of the bow.

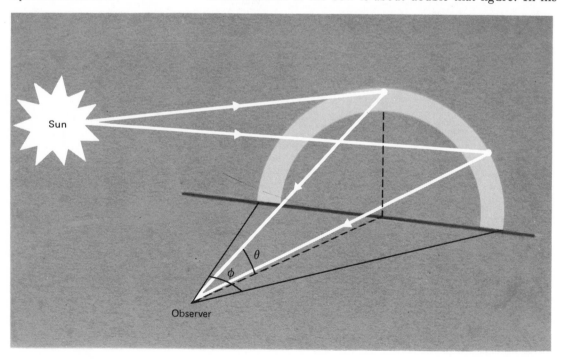

The Causes of the Rainbow

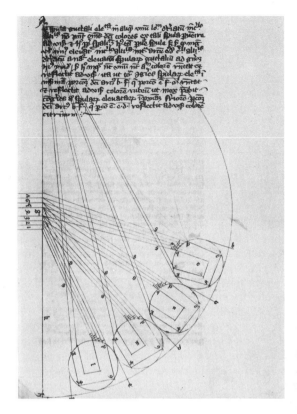

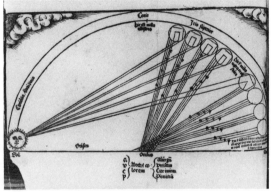

Left Fig. 53: The separation of colours due to the differing elevations of drops. *De Iride*, Basle manuscript, fol. 40.

Below Fig. 54: Rainbow diagram from Jodocus Trutfetter's *Summa in Totam Physicam, hoc est Philosophiam Naturalem*, Erfurt (1514).

account of this work William Wallace suggests one or two possibilities to account for the mistake, but admits himself baffled to explain it convincingly. The trouble is that Theodoric also gives half the correct value for the angular width of the bow in the horizontal plane, the angle ϕ in the diagram. It is possible that he was using incorrectly calibrated instruments, but this hardly seems likely.

The final geometrical problem was to explain why the bow was an arc of a circle. Theodoric's solution depended on noticing that the rainbow, the sun, the raindrop and the observer all lie in one vertical plane. One can imagine that as one pivots this plane about a vertical axis through the observer the illuminated drops capable of reflecting and refracting a specific colour to the eye must be lower and lower, so appearing as a bow.

Rainbow studies after Theodoric

There was not much further systematic work on the rainbow until the time of Descartes. In *Les Météores* of 1637 Descartes gives an account of the physics of the rainbow disconcertingly

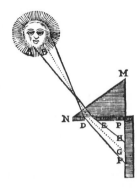

Fig. 55: Descartes's apparatus for the separation of the colours. From the *Discours de la Méthode*, Paris (1668), p. 371.

similar to that of Theodoric, to whom he makes no reference. The similarity extends to the use of glass globes to model raindrops and for tracing the rays of light. One's suspicions are further aroused by the fact that one Jodocus Trutfetter published an account of Theodoric's work on the rainbow, including copies of his diagrams, in 1514. It seems very likely that Trutfetter's or some similar treatment was known to Descartes.

However, Descartes did make a vital and original contribution to the theory of the phenomenon which rounded off the main features of the geometrical optics of the bow. Why is the maximum elevation of the bow 42° or thereabouts? The drops which look red to an observer will be at a different elevation from those which look blue to him, because, as Theodoric had originally established, red and blue rays are differentially refracted. Suppose one were to trace all possible paths of rays of light through a drop, say in a drop that is so positioned that only the red rays reach the eye of the observer. Using Snell's Law of Refraction to guide his calculations Descartes showed that the paths tended to cluster at about 42°. The blue rays in those drops would be refracted at a slightly different angle, and so would miss the eye of an observer, but the blue rays from other drops slightly differently positioned would reach it. These rays would cluster at a slightly different angle. So the bow appears to have a certain breadth and the blue and red parts of it are separated.

Further reading

Grant, E. (ed.), *A Source Book of Medieval Science*, Cambridge, Mass., 1974 (contains a translation of parts of the *De Iride*).

Descartes, R., *Les Météores*, Discours VIII of *Discours de la Méthode et les Essais*, Leyden, 1637.

Boyer, C. B., *The Rainbow: from Myth to Mathematics*, New York, 1959.

Crombie, A. C., *Augustine to Galileo*, New York, 1959, vol. I, pp. 110–11.

Wallace, W. A., *The Scientific Methodology of Theodoric of Freibourg*, Fribourg, Switzerland, 1959.

E

Exploiting an Accident

Systematic studies of phenomena depend on the experimenter having a well-formulated hypothesis, and a clear idea of the phenomena that are to be expected in the experimental procedure. But sometimes accident intervenes, and unexpected and sometimes mysterious results are noticed. Such accidents do sometimes get incorporated into scientific knowledge, but only if the person who runs across them has a theory in terms of which they can readily be interpreted. **Louis Pasteur** was looking for *some* way of attenuating the virulence of an infective agent and found *the* way by accident. **Ernest Rutherford** was not looking for atomic disintegration at all, but he found an unexpected phenomenon that properly interpreted pointed directly to it.

9. Louis Pasteur

The Preparation of Artificial Vaccines

Fig. 56: Louis Pasteur.

Louis Pasteur was born in Dole in the Jura region of France in 1822. His father, after service in one of the crack regiments of Napoleon's army, set up in business as a tanner. Pasteur grew up in Arbois where his father rented a tannery. He had most of his schooling at the Collège d'Arbois, and was rated an indifferent pupil. He seems to have been ambitious for recognition, but determined to acquire it by hard work. He had great difficulty in getting into one of the *Hautes Écoles* in Paris to further his education. He took his baccalaureate at Bescançon and finally entered the *École Normale*. He passed his agrégation in 1846 and took his doctorate in 1847. His high achievement in these examinations led to his being appointed as a laboratory assistant in the *École*.

Pasteur's earliest work was on the optical activity of certain crystalline substances, that is their ability to rotate the plane of polarized light to the right or to the left. He showed experimentally that this power derived from the asymmetrical geometry of the crystals, and surmised that the crystal structure was itself a reflection of molecular asymmetries. In 1848 he was appointed Assistant Professor in Strasbourg, and in 1849 married Marie Laveur, the daughter of the Rector of Strasbourg Academy. In all they had five children, though three died in infancy. He was appointed Professor in 1852. By then he had been internationally honoured for his work on crystallography.

His interest in the biological applications of chemical studies derived in part from a life-long conviction that somehow asymmetry and life were connected manifestations. In 1854 he moved to Lille, and about this time began to develop an interest in the mechanism of fermentation. By generalizing the idea that a yeast was necessary to all fermentation, he came to a germ theory. In 1857 he moved back to Paris as Director of

Scientific Studies at the *École Normale* he had worked so hard to enter.

Once in Paris Pasteur lost no time in cultivating those likely to assist in the financing of research. He formed a fairly close relation with Louis Napoleon (Napoleon III) and his Empress, and courted a certain amount of public disapproval by continuing to speak well of them after their deposition in 1870. In the early 1860s Pasteur became involved in the spontaneous generation controversy, the argument about whether life forms could arise from non-living matter. He used his knowledge of yeasts to demonstrate that the apparent instances of this phenomenon were really caused by air-borne spores. The techniques for studying fermentation 'germs' were also applicable to the study of the causes of disease, and he turned to the investigation of a plague that was damaging the silk-worm industry.

In 1868 he suffered a stroke that led to partial paralysis of his left side. To continue his work he was obliged to employ a strong force of assistants.

The study of diseases, and the promotion of a germ theory of disease corresponding to his germ theory of fermentation became Pasteur's last major area of work. During the Franco-Prussian war of 1870 and the Commune, he remained out of Paris, working on the study of the processes of fermentation involved in wine production. On his return he began to take an increasing interest in the understanding and cure and prevention of human and animal diseases. After his retirement from active teaching in 1874 his attention turned to the popular problem of anthrax. In subsequent work on other more virulent diseases such as rabies, he turned increasingly to the help of assistants, partly because of his revulsion from the necessary vivisections the research required.

In 1886 he suffered a heart attack, and from that time his health steadily declined, with another stroke in 1887 and a final cerebral haemorrhage in 1894 from which he did not really recover. He died in 1895.

Disease theory before Pasteur

As early as 1626 J. B. van Helmont had proposed that diseases should be looked on as the effects of an invasion of the body by an army of alien beings (*archeae*). Once they had established a foothold he supposed that they took over the vital processes of the host for their own benefit, producing waste products that were poisonous to the victim. In essentials this theory anticipated modern ideas. But for more than 200 years it shared the field with a rival, that diseases were malfunctions of the diseased organism, which, roughly speaking, poisoned itself. Some conditions were thought to be the effect of external

Louis Pasteur

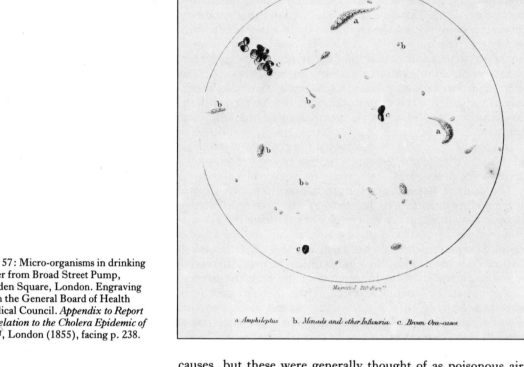

FROM WATER FROM PUMP IN BROAD S.T
GOLDEN SQUARE.

Pl.9.

Magnified 220 diam.

a. *Amphileptus.*   b. *Monads and other Infusoria.*   c. *Brown Ova-cases.*

Fig. 57: Micro-organisms in drinking water from Broad Street Pump, Golden Square, London. Engraving from the General Board of Health Medical Council. *Appendix to Report in Relation to the Cholera Epidemic of 1854*, London (1855), facing p. 238.

causes, but these were generally thought of as poisonous airs (*mal'arie*) rather than alien and hostile organisms.

In the light of the bad-smells-as-causes-of-disease theory some cleaning up of the environment had begun by the beginning of the nineteenth century. The only other prophylactic treatment that had had any real measure of success was vaccination, the preventive for smallpox developed by Edward Jenner. Jenner had supposed that the cowpox, for which the Latin was 'variola vaccinae' (from *vacca*, the cow), was the very same disease entity as human smallpox but of attenuated virulence.

By the mid-nineteenth century there was growing evidence for the association of disease with the presence of micro-organisms. Schwann and others had shown by microscopical studies of various fluids taken from diseased men and animals that there were specific forms of microbes present when the diseases were manifested, but absent in health. The defenders of the old view argued that these microbes were a side-effect of the disorder brought on by the malfunctioning of the body in poor health, coming into being through spontaneous generation.

It should be clear that three steps were needed to break through into the modern conception of disease. First it had to be shown that diseases were the effect of attacks by micro-organisms. But this required that the theory of the spontaneous generation of micro-organisms be finally refuted. And thirdly, the vaccination process of Edward Jenner had to be understood and generalized. To each of these steps Pasteur was the major contributor. But in this section I will describe in detail only one of his contributions, the discovery of the method for the production of vaccines.

Pasteur had devoted a great deal of time and effort to the unravelling of the mechanism of fermentation. He had demonstrated that the presence of a living organism, such as yeast, was the most important factor. Fermentation was really no more than the life process of the specific organism involved in each kind of fermenting. In effect Pasteur established the 'germ' theory of fermentation. Now this, together with his proof that fermentation could not start spontaneously, was readily generalized to a germ theory of disease. Indeed Lister seems, on his own account, to have almost literally seen the putrefaction of wounds as a kind of fermentation. His use of carbolic acid as a disinfectant was a direct application of this idea. Even the anthrax investigations, begun by Davaine, were sparked off by the similarity he noticed between the description of a ferment identified by Pasteur and the rod-like baccilli he had found in the blood of diseased animals.

In order to find one's way around the now unfamiliar terminology of the mid-nineteenth century one must go back to the then novel distinction of viral from microbial diseases. Whatever one believed about the role of microbes in the causation of disease one could make a distinction between the diseases in which they were present, and those in which they were not. In the latter there was some poison or 'virus' responsible. Furthermore, it was viral diseases, smallpox and others like it, which induced immunity; that is, if one survived one attack one could not contract the disease again. Very soon the term 'virus' became generalized to include any disease-causing agent, including microbes. This is how the term is used in contemporary English translations of Pasteur's original papers.

One more puzzling fact must be mentioned in order to understand Pasteur's researches. Medical men knew that the virulence of a disease, whatever its cause, was not always the same. Epidemics came and they went. Diseases occurred in more or less severe forms. The first systematic investigation of variable virulence came in an early study by Pasteur of the septicaemia microbe. He showed that its virulence was very different from different 'cultures', as laboratory preparations of micro-organisms are called. Perhaps, he asked himself, there was something about the cultures which changed the microbe in that way.

The discovery of the attenuation of 'viruses'

In most research efforts it is impossible to isolate a single experiment and locate a great discovery at some one point in an investigation. The study I am about to report centred on two major experimental investigations, the one a study of chicken cholera, the other of anthrax. They are intimately linked, and the final result required both.

Chicken cholera is an epidemic disease of fowls, leading quickly to death. It is accompanied by some very characteristic symptoms, including drowsiness and anoxia, oxygen starvation, shown by the loss of good red colour in the comb. Toussaint had shown that a characteristic microbe was associated with chicken cholera, easily identified in the blood of infected birds. In pursuit of his general thesis that both fermentation and disease were caused by micro-organisms, Pasteur set about an experimental programme to isolate the micro-organism in a pure culture. Then by injecting it into hens he would prove that chicken cholera was caused by the microbe. By using chicken broth as a medium he was able to cultivate the microbe and to show that it maintained its virulence through many successive cultures, new ones being made every day.

In 1879 Pasteur went on a summer holiday to Arbois, his home town, from July to October. He left behind in his laboratory the last of the chicken broth cultures, recently infected with the cholera microbe. When he returned in October the cultures were still there. So he immediately tried to restart the experiment by injecting some of these old cultures into fresh hens. Nothing happened. 'Chance favours only the prepared mind', said Pasteur. It certainly did in this case, since he now decided to restart the programme from the beginning with fresh virulent microbes, with the hens he had already injected with the old cultures. These hens did not develop the disease. Pasteur immediately drew the right conclusion. He had found a way of attenuating the 'virus' artificially.

He was very cagy in his announcement of this discovery. No mention of accident in the following: '. . . by simply changing the process of cultivation of the parasite; by merely placing a longer interval of time between successive seminations, we have obtained a method for decreasing virulence progressively, and finally get at a vaccinal virus which gives rise to a mild disease, and preserves from the deadly disease.'

Several things now needed to be done. First it was necessary to study the effect of successively longer time intervals between preparing new cultures of the microbe from old to try to find out just how much time was required to make the microbe harmless. It turned out that there was a relation between time and decrease of virulence. For intervals of over a month between reseeding cultures no attenuation was observed, but after that the longer the gap the greater the attenuation. To find this out Pasteur had to develop a measure of virulence. This he did by defining the relative virulence of two strains as proportional to the relative numbers of deaths they produce in the same species when the creatures are infected in the same manner and under the same conditions.

Next the mechanism of attenuation needs to be elucidated. Pasteur had long been interested in the role of oxygen in fermentation, and immediately thought of the possibility that the length of time that cultures remained without renewal of microbes or medium would also be a measure of the exposure of the microbe to oxygen. He sealed up some tubes with chicken broth, fresh infections of a virulent strain and a little air, and let them work. After a few days any further

Fig. 58: Pasteur in his laboratory.

development stopped. Similar cultures were prepared in open flasks. Even after two months, by which time the culture in the open flasks had become completely innocuous, when he opened one of the sealed tubes and used that culture, long since quiescent, to infect birds, the culture proved to be of a 'virulence of the same degree as that of the liquid which served to fill up the tube. As to the cultivations open to the air, they were found either dead or in a condition of feebler virulence.'

But what had happened to the microbes to make them so feeble? Pasteur was unable to find out. 'If any such relations [between morphological distinctions and between forms of different virulence] sometimes appear, they disappear again to the eye working through a microscope, on account of the extreme minuteness of the virus.'

The relation of vaccine to disease virus was now clear. '. . . for while discussions continue on the relations of vaccine to [smallpox] we possess the assurance that the attenuated virus of chicken cholera is derived from the very virulent virus proper to this disease, that we may pass directly from one form of the virus to the other. The fundamental nature of each is the same.'

While the discovery of attenuation depended on a combination of prepared mind and happy accident, the subsequent investigations were perfectly Baconian. Time is associated with attenuation, but what is the 'latent process' of which the time factor is the outward manifestation? Pasteur never did satisfactorily answer that question.

Subsequent development

The story of the generalization of these results and the creation of practical vaccines for diseases afflicting man is unusual, since it was Pasteur himself who was the prime worker in this.

From a scientific point of view the two most important of his subsequent pieces of work were the development of an anthrax vaccine, and his discovery of how the disease was spread, and the dramatic results of his later work on rabies.

The remarkable thing about these researches is the way theory guided Pasteur through a thicket of confusing empirical difficulties. He was quite clear that, from the point of view of the biology of the micro-organisms, the host was just another environment. There was nothing special about the distinction between chicken broth as a medium for the culturing of cholera microbes and chickens. In both the microbes grew and flourished. So different species of animals could be thought of as possible sites for attenuation of 'viruses'.

Anthrax was known to be associated with a microbe, but the discoverer of this fact, Toussaint, had mistakenly tried to develop a purely chemical vaccine by filtering out the mi-

crobes. By an ingenious experiment involving chilling hens,
Pasteur showed that the disease symptoms were not caused by
the chemical by-products of the activity of the microbe in a
culture, but by the micro-organism itself. The difficulty of
attenuating the anthrax bacillus came about because it readily
protected itself from excess oxygen, heat and so on, by forming
resistant spores. But Pasteur found that by careful control of
the heating of his culture he could prevent the formation of
spores. Between 42°C and 44°C spores were not formed, but
any error was fatal since at 45°C the microbe died. However,
the results were very satisfactory. Time worked again, and
after only eight days full attenuation had taken place. To test
all this in the kind of glare of publicity Pasteur loved, the great
Pouilly-le-Fort test was arranged.

A. M. Rossignol, a one-time critic of Pasteur, undertook the
organization. On 5 May 1881, twenty-four sheep, one goat and
six cows were injected with an attenuated anthrax strain. On 31
May a fully virulent culture was injected into all thirty-one
vaccinated animals and twenty-nine unvaccinated. By 2 June
all the vaccinated animals were still healthy, while by the
evening of that day all the unvaccinated sheep were dead and
the unvaccinated cows very ill.

The Preparation
of Artificial Vaccines

Fig. 59: The great anthrax experiment
at Pouilly le Fort. Contemporary
woodcut by A. Gançon. Musée
Pasteur, Paris.

The result was a triumph for Pasteur. But though the process spread rapidly throughout France and England, and Pasteur's own 'factory' manufactured the vaccine in great quantities, he was subjected to a jealous and spiteful attack from his German rival Robert Koch, mortified by the evident success of Pasteur's work. Only agitation by the German farmers finally persuaded the German Ministry of Agriculture to introduce the vaccine.

But rabies was not only a much more dangerous disease. It was, as we now know, caused by a virus, in our sense of that word. So there was no chance of microscopical identification of the organism to serve as the stock for culturing a weaker strain. But Pasteur had noticed one important thing. The disease was primarily an attack on the nervous system, and was clearly identifiable in the brains of its victims. Returning now to his fundamental idea of animals as biological environments, he decided to use spinal chord as the culture medium. By infecting rabbits he was able to obtain rabbit spinal chords which were infested with the mysterious micro-organism. These were hung up in sterile atmospheres and slowly dried. As they did so the rabid effect of injecting animals with a paste made from strips of the chord became weaker and weaker. Again attenuation was just a matter of time, but in the right medium. Eventually, in a legendary case, Pasteur was persuaded to try the vaccine on a child that had been bitten by a rabid dog, and that child survived.

Further reading

Pasteur, J. J., 'Attenuation of the Virus of Chicken Cholera', *Chemical News*, 43, 1881, pp. 179–80 (translation of the paper originally appearing in the *Comptes rendus ... de l'Académie des Sciences*, 91, 1880).

Suzor, J. R., *Hydrophobia: an Account of M. Pasteur's System*, London, 1887.

Cuny, H., *Louis Pasteur: The Man and his Theories*, London, 1965.

Dubos, R., *Pasteur and Modern Science*, New York, 1960.

Winner, H. I., *Louis Pasteur and Microbiology*, London, 1974.

10. Ernest Rutherford

The Artificial Transmutation of the Elements

Ernest Rutherford was born of a Scottish father and an English mother in Nelson, New Zealand, in 1871. His father was a small farmer and something of a general engineer, and his mother was a schoolteacher. He won a scholarship to Nelson College for his secondary education. He excelled at school, particularly in mathematics. Another scholarship took him to Canterbury College at Christchurch, then one of the constituent colleges of the University of New Zealand, in 1889. He took his M.A. in 1893 with a double First in Mathematics and Mathematical Physics. He had already begun research work into magnetism, and in 1894 to 1895 he developed a detector for radio waves.

In 1895 he was awarded an 1851 Exhibition Scholarship to Cambridge, where he worked under J. J. Thomson, in the Cavendish Laboratory. His first studies in Cambridge were in collaboration with Thomson, on the ionization effects of X-rays. Then, in 1898, he turned to the exploration of the phenomenon of radioactivity, the emission of radiation from the natural breakdown of elementary substances.

Fig. 60: Ernest Rutherford.

He was offered the chair of physics in McGill University in Montreal in 1898. Not only did this move give him a laboratory of his own, but put him in the financial position to marry Mary Newton, to whom he had become engaged while at Christchurch. Here he began the astonishingly fruitful collaboration with the eccentric Frederick Soddy, who supplied the necessary chemical expertise, in their joint investigation of the properties of radioactive materials. With Soddy, Rutherford formulated the atomic disintegration theory of radioactivity in 1902. He was elected a Fellow of the Royal Society in 1903 and awarded the Rumford Medal in 1904.

In 1907 he returned to Britain as Professor of Physics at the University of Manchester. He immediately attracted around him a group of very talented younger men. He was awarded the Nobel Prize for chemistry in 1908. In 1909, in collaboration with Geiger and Marsden, he carried out the experiments that suggested that atoms consisted of heavy nuclei surrounded by orbiting electrons. At first this discovery was not widely recognized, but it began a very fruitful period of collaboration between Neils Bohr and Rutherford, in the course of which Bohr sketched out the quantum theory of fundamental particles and their interactions.

During the First World War Rutherford worked on problems of submarine detection, but at the same time managed to continue his major researches. The discovery of the artificial disintegration of elements and their forced transmutation came in 1919, the experiment to be described in this section. In 1919 Rutherford finally returned to Cambridge, succeeding J. J. Thomson as Director of the Cavendish Laboratory. Here he worked with Chadwick on systematic studies of the artificial disintegration of the elements, and it was here that with Oliphant and Hunter he produced the first nuclear fusion, the creation of atoms of a heavier element by fusing the atoms of a lighter one.

He was awarded the Order of Merit in 1925 and elevated to the Peerage in 1931. He died in Cambridge in 1937.

The state of knowledge before Rutherford's experiment

In trying to set out the history of the problem of the transmutation of the elements a great deal depends on what one takes the term 'elements' to mean. In antiquity the distinction between compounds and elements, as we know it, was not clearly drawn. Historically the most important distinction was rather between the metals and non-metals. But there was a doctrine of elements. They were thought of as the basic principles which, in combination, formed the familiar substances of the earth's crust, such as metals, organic materials, stones and so on. There were generally thought to be four of these elements or principles, which existed in different proportions in different substances. The basic principles were indestructible, and most scientists of antiquity thought that they could not be transformed into one another.

However, from the time of Alexandria's dominance of the scientific culture of the mediterranean world, say from about 300 BC, a growing number of students of nature came to think that ordinary substances could be transformed into one another. Food could be transformed into flesh, ice into water, ore into metal and so on. This idea, well grounded in common experience, was generalized to include all substances, including the metals. So it was thought that by suitable manipulations lead or tin could be transmuted into gold, iron or anything else. This project had a double significance. By being

Fig. 61: *The Alchemist*, by Thomas Wijk (1616–77). Collection of Dr Alfred
Bader.

connected with the signs of the Zodiac, the metals had been related to astrological theories and were thought to have powers of a rather special kind. So gold, as the supposedly most perfect metal, began to assume an importance over and above its role in the economic systems of the time. To find a way of transmuting common metals into gold would then not only be of some economic advantage (even in the ancient world not everyone had fully grasped the folly of inflation), but it would also open up the technical possibility of creating other perfect substances, for instance the perfect medicine, the *panacea*.

Chemists, in this tradition, believed that the metals, like all other substances, were formed from different proportions of the four basic elements. They supposed that if they could find out the proportions in baser substances they could add to or take away from the amounts of the elements which were out of balance, so to speak, and so modify the substance. If they could hit on the perfect balance, then they would have created gold. There were mathematical theories derived from some of the simple properties of natural number sequences, such as magic squares, which suggested that some proportions were well grounded mathematically. The research programme based on these theories, which we call 'alchemy', was a total failure. But in the course of trying to do the impossible, alchemists discovered a great many useful chemical reactions and preparations.

Some time between the Renaissance and the end of the nineteenth century, the whole idea of transmuting the elements, now thought of as the most elementary amongst the ordinary substances of nature, had fallen into disrepute. The exact story of the development and spread of this opinion is not really known. The scientists of the seventeenth century, who, like Robert Boyle, believed that material substances were made of different structures of basically similar corpuscles, had no difficulty with the idea of transmutation, though they had a lordly contempt for alchemical theories and attempts at transmutation based upon them. After all, if Boyle had been able to break down a substance into an undifferentiated broth of its basic constituents and recombine them in a different arrangement, he would have transmuted one substance into another. Whatever the history of the dogma that transubstantiation was impossible, it was very well entrenched by the end of the nineteenth century.

Rutherford's first contribution to the business was his theory of radioactivity. Becquerel had noticed that some minerals gave out 'rays' which made the mineral fluorescent, and which would blacken photographic plates. The Curies were isolating the active constituents from the mineral. But the theory of radioactivity was primitive. Most thought it a form of fluores-

Fig. 62: Marie and Pierre Curie
pictured in their laboratory.

cence, a chemical reaction producing light. Only Rutherford
seemed to have grasped how fundamental a process was going
on when a radioactive substance gave out rays. He proposed
the novel hypothesis that the source of the rays was a
disintegration of the very atoms themselves. Disintegration
would lead to the fission of the atom into smaller atoms, which
would necessarily be atoms of a different substance. This idea
explained why radium, a heavy metal, decayed through various
intermediate substances into lead, a somewhat lighter metal
than radium, atom for atom. So far as anyone knew, this
disintegration was confined to the very heavy elements, and it
was spontaneous.

The first artificial transmutation of an element

Ernest Rutherford

There is a widespread myth that scientists do experiments to test hypotheses. This implies that a hypothesis is first formulated, somehow, and then various consequences are drawn from it which are put to the test. If they fail, the hypothesis is to be rejected, and if they pass, then it can be accepted for the moment as plausible. Rutherford's experiment is not like this. It was not a test at all. The experiment, designed originally to explore the strength of the impetus imparted to products of the disintegration of atoms, revealed a surprising effect. Rutherford had the wit to formulate a hypothesis to fit the unexpected effect; a hypothesis which developed directly out of and extended the ideas he had already introduced to explain natural radioactivity.

The original experiment involved the use of a source of α-particles, common products of the disintegration of heavy atoms (now known to be nuclei of helium atoms), a chamber into which different gases of different stopping powers could be introduced, and a screen which would detect particles which had either passed right through the gas in the chamber or had been emitted by collisions between α-particles and molecules of the enclosed gas.

When this apparatus is equipped with a Radium-C source to produce α-particles and filled with air, there appear 'scintillations on the screen far beyond the range of the α-particles' emitted at the source. At first sight they seemed to Rutherford similar to the 'swift H [hydrogen] atoms produced by passing α-particles through hydrogen'. When an α-particle hits a hydrogen atom it gives it a 'shove', projecting it with very high

Fig. 63: The experimental arrangement.

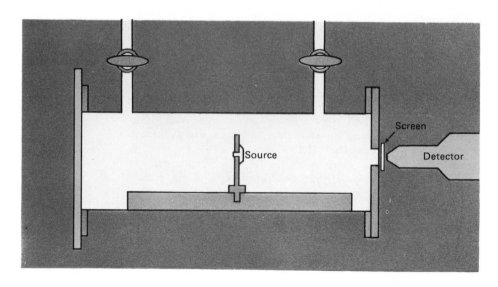

velocity and long range. When oxygen or carbon dioxide (i.e. constituents of air other than nitrogen) were introduced into the apparatus the scintillations due to long-range particles were much reduced. 'A surprising effect was noticed, however, when dried air was introduced.' Instead of the number of long-range scintillations being reduced it was increased by a large amount, and what is more, they were of very long range (19 cm). To what could they be due?

Rutherford began a long series of subsidiary experiments to answer this question. Each experiment was designed to eliminate a possible source for the mysterious fast particle. There are some swift oxygen and nitrogen atoms produced by collisions with α-particles, but these have a range of about 9 cm only. By using a screen between the chamber and the detector screen, which had a stopping power greater than 9 cm of air, 'these atoms are not completely stopped'. He showed that the anomalous effect was not due to water vapour since it still occurred with carefully dried air. It was not due to dust particles since carefully filtered air produced the same effect. But if it were due to the nitrogen of the air, in some way, then the long-range particles should continue to be produced and perhaps even increase, if nitrogen from some chemical source was introduced. And that is exactly what happened. The increase was precisely what would be expected as the amount of nitrogen is increased from 80 per cent as in atmospherical air, to 100 per cent.

'The results so obtained show that the long-range scintillations obtained from air must be ascribed to nitrogen.' But the next step was to show that they are due to collisions with α-particles. This could be presumed if there were any evidence

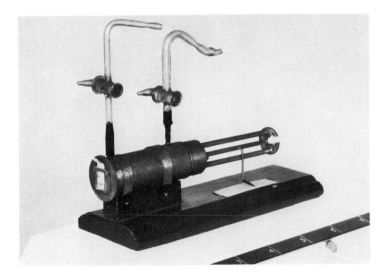

Fig. 64: Rutherford's apparatus, now in the Science Museum, London.

that they were due 'to collisions of α-particles with atoms of nitrogen throughout the volume of the gas'. One obvious test would be to change the pressure of the gas. If the number of scintillations decreased directly proportional to the decrease in the pressure of the gas then this would be good evidence. Further, Rutherford showed that the range of the expelled atom that produced the scintillation was proportional to the range of the expelling atom. When a target molecule was hit which was further from the source of α-particles, the expelled particle would go correspondingly further. This simple reasoning clearly implied that the effect must have been due to collisions.

The hypothesis–test stage, so often proffered as the whole of science, comes last in this piece of work. All the evidence derived from common-sense examination of the effect points to the theory that the long-range scintillations are due to hydrogen atoms, and *that they must come from within nitrogen atoms*, as the product of an artificial disintegration, leaving behind transmuted atoms of a lighter element.

But first one must be sure that the long-range particles are indeed hydrogen atoms. The degree to which a moving particle

Fig. 65: A radioactivity kit, a curious practical application of a newly discovered branch of science. Museum of the History of Science, Oxford University.

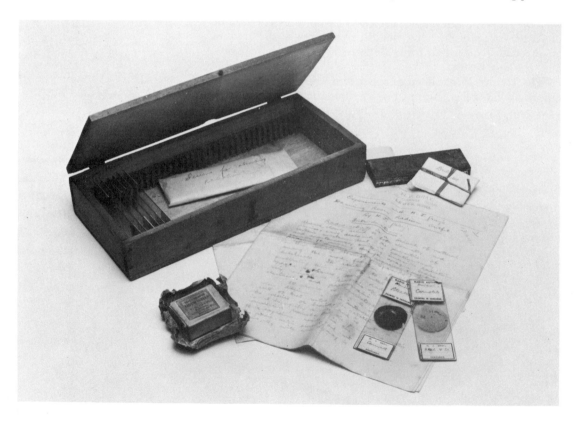

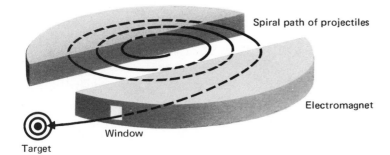

Spiral path of projectiles

Electromagnet

Window

Target

Fig. 66: The path of a beam in an accelerator.

is deflected in a magnetic field is proportional to the ratio of charge to mass, a very characteristic ratio, for particular kinds of atoms. So if the long-range particles were deflected by the same amount as hydrogen atoms similarly prepared, and by the same magnetic field, it could be presumed that they were indeed hydrogen atoms. Rutherford did get an effect of the right order, but the 'numbers involved were too small' for him to be satisfied with the experiment as a proof. But everything added up to the near certainty that that was what the long-range particles were. Now for the interpretation of the experiment.

'... we must conclude', says Rutherford (p. 586), 'that the nitrogen atom is disintegrated under the intense forces developed in close collision with a swift α-particle, and that the hydrogen atom which is liberated formed a constituent part of the nitrogen nucleus.' But this interpretation must not be just fudged up *ad hoc* to explain the effect – it must be able to be seen as a natural extension of theories already well established. Rutherford goes on to show how the new effect fits in. 'Considering the enormous intensity of the forces brought into play, it is not so much a matter of surprise that the nitrogen atom should suffer disintegration as that the α-particle itself escapes disruption into its constituents.' From nitrogen Rutherford had been able to produce another element, hydrogen, by his active manipulation of the necessary equipment.

With characteristic prescience he went on to suggest the research programme that has dominated nuclear physics ever since. '... if α-particles – or similar projectiles – of still greater energy were available for experiment, we might expect to break down the nuclear structure of many of the lighter atoms.'

Nuclear research after Rutherford

This programme was able to be fulfilled in two ways. By the 1930s a new kind of ray had been discovered, the cosmic ray. There were particles coming into our atmosphere from outer space with immense energies. By exposing photographic plates

Ernest Rutherford

to the cosmic radiation in balloons rising high above the heavier, denser parts of the atmosphere it was possible to record collisions between these very powerful projectiles and atoms of the air. All kinds of new particles, supposedly constituents of ordinary atoms, have been found. But this technique, through fruitful when it comes off, is quite happen-chance. One may wait about quite a while for a cosmic ray collision, and when it comes one may not have a collision involving the kind of energies which one wants.

The second line of development was the design and construction of artificial accelerators. The basic principle is simple. A charged particle like an electron or a proton is

Fig. 67: Rutherford's research room in the Cavendish Laboratory, Cambridge.

attracted by an electric field of suitable polarity, and so is accelerated. By arranging a sequence of electric field generators, which are switched off as the particles pass into the grasp of the next field, huge accelerations are achieved. To make the equipment more compact, another effect is often employed. In a magnetic field charged particles tend to be drawn into a curved path. By arranging a suitable magnetic field a stream of particles can be made to go in a spiral or circular path; with electric fields to accelerate them particles can be made to go round and round, up to a million times, within the circular core of the apparatus until they are discharged at a target with terrific energy. With these machines the exploration of the

The Artificial
Transmutation
of the Elements

Fig. 68: Inside the CERN accelerator tunnel at Geneva, showing the curving line of magnets. The circumference of the circle is 6.9 km.

Ernest Rutherford

structure of the nucleus of the atom has been carried on with great success in recent years.

It is sometimes argued that science does not accumulate knowledge but lurches from one world view to another. This can hardly be true in the short term. Rutherford's experiments and his interpretation of the anomalous effects as the disintegration of complex atomic structures by collision with projectiles depend on his general acceptance of Thomson's interpretation of the phenomena of gas discharge. Without the idea that there are subatomic electrically charged material projectiles, Rutherford could have not made his discovery.

Further reading

Rutherford, E., 'The Collision of α-particles with Light Atoms', *Philosophical Magazine*, 6th series, vol. 37, 1919.

Andrade, E. N. da C., *Rutherford and the Nature of the Atom*, London, 1964.
Schonland, B., *The Atomists, 1805–1933*, Oxford, 1968, ch. 7.

F

Null Results

What if an experimental procedure is carried through and there is no result at all? One of the most famous of all experiments, that of **A. A. Michelson** and **E. W. Morley**, involved a manipulation that had no measurable effect. If this happenes there are two possible explanations to hand. The manipulation did produce an effect but there was a reciprocal effect too, which 'cancelled' the first one out, so to speak. Sometimes experiments contrive to test a hypothesis by deliberately producing such a cancellation. But it might be that the theory in terms of which the result aimed at was to be expected is not just wrong, but somehow conceptually incoherent. In the case of the experiment to be described in this section both kinds of explanation were tried out. Eventually one of the latter sort came to be accepted as the most fruitful way to account for the null result.

11. A. A. Michelson and E. W. Morley

The Impossibility of Detecting the Motion of the Earth

Fig. 69: A. A. Michelson.

Albert Abraham Michelson was born at Strelno in Prussia in 1852. His father seems to have been something of an adventurer. When Michelson was still an infant the family migrated to the United States. Michelson's father set up in business trading with the gold miners of Nevada. The child was brought up in Virginia City, a classic 'gold rush' town. He was boarded away from home for his later schooling, in San Francisco. Proving exceptionally able he completed his higher education at the U.S. Naval Academy at Annapolis, though he had some troubles fulfilling the entry regulations. After a tour of seagoing duty in the Navy he returned to the Academy as an instructor in the physical sciences. In 1877 he married Margaret Heminway, and so acquired a very wealthy father-in-law.

His interest in the measurement of the velocity of light seems to have begun about 1878, and his first experiments were conducted in that year, with apparatus paid for by his father-in-law. From 1880 to 1882 he did postgraduate study at several European universities, notably with Helmholtz in Berlin. There he began to study the optical effects of interference between light waves, with equipment paid for by Alexander Graham Bell, the telephone engineer.

Having resigned from active naval duty in 1881, he returned to the United States in 1882 and joined the Case School of Applied Science in Cleveland. By then Morley had begun his work at the neighbouring Western University. The famous experiment was done in 1887.

In 1889 Michelson moved to Clark University, and then in 1893 to Chicago to head the physics department of the new university. By this time his interests had shifted from the problem of the velocity of light to other uses of interferometry, such as the standardization of the metre against the wavelength of cadmium light.

Having divorced his first wife, he married Edna Stanton in 1899. He returned to the Navy during the First World War as a Reserve officer, and worked on the development of optical range finders. After the war he spent more and more time in California. This was partly for pleasure but he was also working on some new projects, including measuring the diameter of the stars. Once again the velocity of light absorbed his interest, and he began working with apparatus set up on adjacent mountain peaks, giving much longer distances between source and detector.

Michelson was much honoured in his life time. He received the Copley Medal of the Royal Society, and was the first American citizen to be awarded a Nobel Prize. He died in Pasadena in 1931.

Edward Williams Morley was the other member of the famous team. He could hardly have come from a more different background. He was born to strict Congregationalist parents in Newark, New Jersey, then a country town, in 1838. His father was a minister in the Church, and Edward was schooled at home by his parents. Not till he was nineteen did he have any public education, when he went to Williams, his father's college. He was intended for the ministry, and after taking his first degree began studies at the Andover Theological Seminary in 1861.

He began his teaching career at the South Berkshire Academy in 1866. It seems that he taught both theology and general science. In 1868 he married Isabella Birdsall, and in the same year became minister to the Congregational Church in Twinsburg, Ohio. Western Reserve College was nearby at that time, and he was invited to do some teaching there. When the college transferred to Cleveland in 1882, as Western Reserve University, Morley went with it as professor of chemistry.

Fig. 70: E. W. Morley.

Morley had a life-long obsession with accuracy. His style was something like that of an American Berzelius. He worked on chemical and meteorological problems that centred on very accurate determinations of the physical properties of oxygen. His chemical interests drew him to spectroscopy and the study of interference phenomena, and this led to the collaboration with the worldly and ebullient Michelson.

In 1906 Morley retired from Western Reserve and returned east to West Hartford, Connecticut, where he died in 1923.

The situation prior to the Michelson–Morley experiment

The experiment seems at first glance to be concerned with the measurement of the velocity of light, but its significance is very much deeper. The idea that light must have a finite velocity seems to have been incorporated, so to speak, in the material

A. A. Michelson
and E. W. Morley

Fig. 71: Fizeau's wheel. A beam of light passes through the teeth of the fast-rotating toothed wheel $a' - a$ (inset labelled fig. 694). Reflected light can reach the observer only when a gap between teeth is opposite both mirror and eye. At the correct rate of rotation a steady light is seen. From the time taken for a tooth to pass, the velocity of light can be calculated. Illustrated in J. C. Jamin, *Cours de Physique*, Paris, vol. 3 (1866), plate 3.

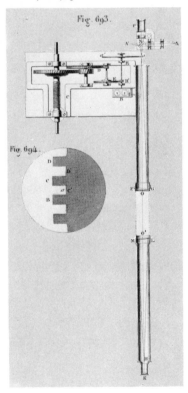

theories of light of Descartes and Newton, since both conceived of light as a stream of particles. Evidence that light has a definite velocity came originally from astronomical observations. Olaus Roemer noticed that the intervals between the eclipses of the moons of Jupiter were different when the planet and the Earth were approaching each other from when they were receding from each other. A simple explanation of this effect is that the light, having a definite and finite velocity, takes a shorter time to reach the Earth when the planets are approaching each other, and a longer time when they are receding. These observations were made in 1675.

Ideally the velocity of light ought to be measured in an earthly laboratory with all the precautions against disturbances that make for a reliable result. No really satisfactory method was available until 1849, when H. L. Fizeau devised an ingenious and simple method, easily set up in a laboratory. The results were consistent with the velocity calculated on the basis of Roemer's observations.

What was light? The particle theory of Descartes and Newton had slowly been replaced by a wave theory. Light was thought of as a transverse vibration in a universal medium, the ether. The ether was supposed to permeate the whole universe, and to be the stationary background to all motions. To grasp how this idea relates to Newton's conception of a mechanical universe we must notice a peculiarity of his famous Laws. They have an important mathematical property, called Galilean Invariance. This property means that Newton's Laws of Motion are the same for all bodies, no matter how fast they are moving relative to each other or to the imagined stationary ether. It follows that there is no mechanical way of detecting one's absolute motion.

If Newton's Laws are the same whatever the relative motions of the systems to which they are applied, there is no way they could be used to determine whether any system of bodies, say our own galaxy, is really at rest. But if light is due to the spread of vibrations in a stationary ether, then this ether might do as a fixed background against which to measure all motions.

The experiment

Imagine a light pulse sent out from a source in the direction of motion of the source. If the light pulse is transmitted by the stationary ether its velocity will be always the same, regardless of the motion of its source or of a detector. Its velocity is constant relative to the *ether*, not its sources and detectors. Now imagine a light pulse sent out at right angles to the direction of motion of a source, and with a detector the same distance from the source as the one for the first pulse. We now have a set-up as in the diagram below. If the source and detectors are imagined to be rigidly bolted to a frame and so are all moving through the *stationary* ether, and both light pulses are moving with the same definite velocity in that ether and relative to it, then the time taken by light pulse 1 to reach the detector 1 will be longer than that required for light pulse 2 to reach detector 2. Detector 1 will have moved on ahead while the pulse of light is moving through the ether at its fixed speed. L1, the length of the rigid bar holding the source and Detector 1 together, will seem to be longer than L2, the length of the bar holding Detector 2. The apparatus designed by Michelson and Morley is essentially a more elaborate version of the equipment sketched in the diagram. Their apparatus seems more complicated because they had to include some way of detecting the minute differences in time of travel of the light in the two directions. Their method for this involved reflecting the light back along the lines of the original pulses and comparing the times by interference effects.

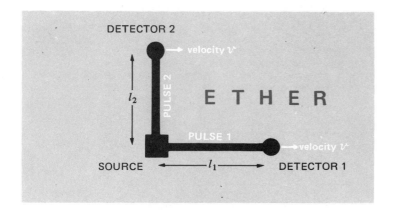

Fig. 72: The Michelson-Morley experiment in the classical world picture.

It is clear that theoretically there should be a difference in the times taken by a pulse of light to traverse the arms of the apparatus. But how could such a very minute difference be detected? The trick was to use the phenomenon of interference fringes. This phenomenon comes about because in one aspect of its behaviour light is like a wave motion. With the wave picture in mind we would expect to detect interference effects when two waves combine (see fig. 73). If the waves A and B combine as in Condition 1 troughs and crests coincide to give an amplified effect. But in Condition 2 troughs and crests will cancel each other out, giving darkness. Suppose A and B come from the same source, but reach the point where they start to recombine by different routes. If the paths of A and B differ by exactly one wavelength we would expect to get Condition 1, but if they differed by half a wavelength we would get

128

A. A. Michelson
and E. W. Morley

Condition 2. Suppose however that we start with Condition 1 and then contrive to make the path of A just a very little shorter than that of B. The crests of B would now arrive before the crests of A. Then the highest point of the combination of the crests would be shifted slightly to the left. By this 'shift' an observer could tell whether the path A had changed in length.

White light is a mixture of lights of different wave-lengths, and for each wave-length there is a different colour (though in practice the eye cannot discriminate to the extent that is theoretically possible). Because of the mixture of waves, when interference effects are studied with white light the bright peaks of full combination of waves are surrounded by coloured fringes. It was to the detection of shifts in these fringes that the Michelson–Morley experiment was directed.

Michelson's first attempt to measure the motion of the earth through the ether took place in 1881. But he had overlooked 'the effect of the motion of the earth through the ether on the path of a ray of light at right angles to this motion'. The

Fig. 73: Shift in the interference fringes: explanatory diagram.

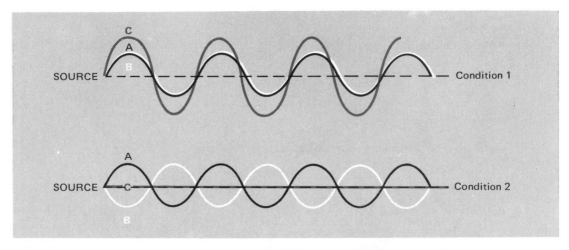

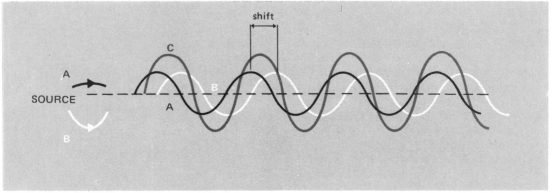

experiment of 1887, with Morley, took account of this effect.

The whole apparatus is assumed to be moving with the earth through the ether. Simple mathematical analysis shows that the difference between the observed lengths of the paths of light in the two perpendicular directions should be Dv^2/c^2, where D is the length of the arm of the apparatus, v is the velocity of the earth through the ether, and c is the velocity of light. As Michelson and Morley point out, 'only the orbital motion of the earth is considered. If this is combined with the motion of the solar system, concerning which but little is known with certainty, the result would have to be modified; and it is just possible that the resultant velocity at the time of the observations was small though the chances are much against it. The experiment will be repeated at intervals of three months, and thus all uncertainty will be avoided.'

If the apparatus is rotated through 90° the difference will still be Dv^2/c^2, but since the arm which determined the longer path will now determine the shorter, the total difference will be $Dv^2/c^2 \times 2$. The effect of this, for an apparatus the size of that used by Michelson and Morley, should be a displacement of the interference fringes produced by the recombination of the beams of light by 0.04 of a fringe distance from their relative positions as observed in the original condition.

The physical principles involved in the experiment are quite straightforward and the observations of no great difficulty. The skill of the experimenters was expended in the refinement of the equipment to eliminate as many sources of error as possible. Two main problems beset them. Stray vibrations could upset the optical part of the apparatus, making the fringes unclear. The shortness of the paths of the rays of light

Fig. 74: The apparatus of 1881, illustrated in Michelson's early paper, 'The Relative Motion of the Earth and the Luminiferous Ether', in the *American Journal of Sciences*, 3rd series, vol. 22 (1881), pp. 120–9.

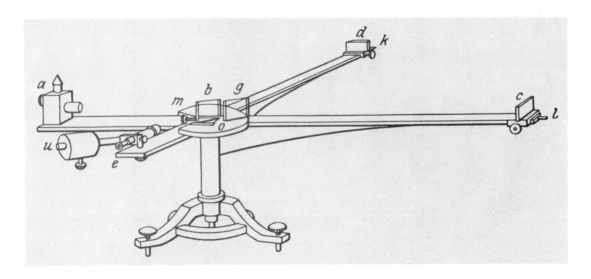

A. A. Michelson
and E. W. Morley

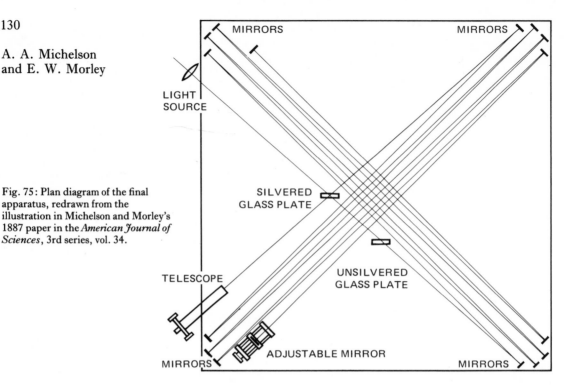

MIRRORS   MIRRORS

LIGHT
SOURCE

SILVERED
GLASS PLATE

UNSILVERED
GLASS PLATE

TELESCOPE

ADJUSTABLE MIRROR

MIRRORS   MIRRORS

Fig. 75: Plan diagram of the final
apparatus, redrawn from the
illustration in Michelson and Morley's
1887 paper in the *American Journal of
Sciences*, 3rd series, vol. 34.

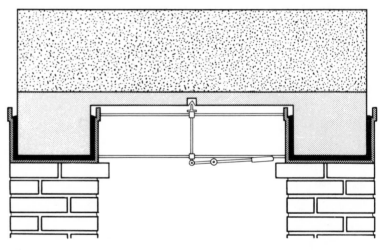

Fig. 76: Cross-section of the final
apparatus, redrawn from the
illustration in Michelson and Morley's
1887 paper.

meant that the effect was small and so difficult to detect with
certainty. Both difficulties were overcome in the final arrangement.

Instead of moving the apparatus to the second position, and
then waiting for it to settle, Michelson and Morley found that

they had much clearer optical effects and less distortion if they rotated the whole apparatus very slowly and steadily in its bath of mercury. They were able to study the fringes while the stone slowly turned without being troubled by distortions. They performed two sets of observations each day, one set at noon, and the other at six in the evening. In this way they hoped to eliminate any effects due to daily changes in the weather. For the noon observation the stone was rotated anticlockwise, and in the opposite sense for the evening.

If we return to the diagram in which the basic structure of the experiment is laid out, we can see that the failure to detect a difference in the length of the paths of the light pulses in the two arms of the apparatus deals a fatal blow to the idea of using the ether as a stationary background against which to measure the 'real' motion of the earth.

But how was the result of the experiment to be explained? Perhaps there is a compensatory change. The apparatus is assumed to be rigid. But suppose that it contracted along the

The Impossibility
of Detecting
the Motion of the Earth

Fig. 77: A contemporary photograph of the final apparatus.

A. A. Michelson
and E. W. Morley

direction of motion, squeezed up by the ether, by just the right amount to compensate for distance the Detector 1 had moved? This would explain why Michelson and Morley failed to detect any change in the fringes. This was Lorentz's solution, and has been called the 'Fitzgerald–Lorentz' contraction after the two men who proposed it.

A more radical explanation involves abandoning the underlying picture of the physical structure of the universe assumed in the design of the experiment, in particular giving up the idea of the ether. If there were no ether there would be no foundation for expecting the result. We could accept the result at its face value. If light had the same velocity, not only with respect to the imaginary ether, but with respect to *whatever* real detector it was measured by, we would expect no positive result.

Fig. 78: Photograph of a fringe shift, taken from an article by D. C. Miller in *Reviews of Modern Physics*, vol. 5 (1933), fig. 7.

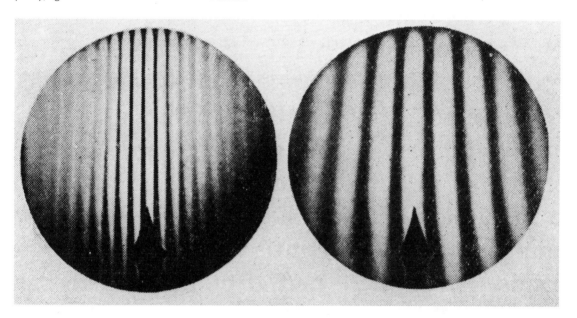

After the Michelson–Morley experiment

H. Lorentz was in close touch with Michelson and Morley, and it was to his theory that the experiment was first considered relevant. He took the view that the apparatus must have contracted in the direction of motion, the former of the above alternatives. This contraction could be explained by a quite natural assumption about the forces that hold the molecules together in the material of which the apparatus was made. Lorentz asks us to suppose that intra-molecular forces are transmitted by the ether just as magnetic and electrical forces are. Any body moving relative to the ether will be subject to a force just as the electric bodies, electrons, are. The force is given by the law

$$\text{Force} = e(E + \text{v}/cB)$$

where e is the electric charge, E represents the electric field, B the magnetic field, v the velocity of the moving body relative to the ether, and c the velocity of light. The faster a body moves relative to the ether the greater the force it will experience, causing it to contract along the line of motion. So, according to Lorentz, the Michelson–Morley experiment gives a null result because the arm of the apparatus that is laid along the direction of motion of the earth as it moves through the ether, has contracted just enough to compensate exactly for the time taken by the light pulse in that direction. As Lorentz himself put it, 'surprising as this hypothesis may appear at first sight, yet we shall have to admit that it is by no means far-fetched, as soon as we assume that molecular forces are also transmitted through the ether, like the electric and magnetic forces of which we are able at the present time to make this assertion definitely. If they are so transmitted, the translation will very probably affect the action between the two molecules or atoms in a manner resembling the attraction or repulsion between charged particles. Now since the form and dimensions of a solid body are ultimately conditioned by the intensity of molecular actions, there cannot fail to be a change of dimensions as well.'

An alternative explanation of the null result follows from Einstein's reformulation of the basic laws of physics. It should be remembered that his work was independent of the Michelson – Morley experiment. Einstein believed deeply in the ultimate unity of the physical world and the simplicity of its fundamental processes. Suppose that *all* the laws of nature were the same for all systems of bodies, including the laws of electromagnetism, the exceptions in the old physics. The old physics had taken Newton's Laws for granted. But suppose physics were to be thoroughly revamped, starting with the electromagnetic laws and presuming that these laws, as they had been formulated by Maxwell, were to be the same for all systems of bodies in any kind of relative motion. Then just as in the old physics there could not be mechanical tests for 'real' motion, so the new physics would imply that there could be no electromagnetic tests. In particular the Michelson–Morley experiment could never work. To complete the programme Newton's Laws would have to be changed so that they would be the same for all systems of bodies, no matter how they moved, according to the rules for electromagnetic laws. If Newton's Laws, invariant under the old Galilean transformation, could be altered to be invariant under a new mathematical condition, one that was tailor-made for Maxwell's electro-magnetic laws, then a perfect unity could be achieved. This new harmony was achieved in the Special Theory of Relativity.

Further reading

Michelson, A. A., and Morley, E. W., 'The Motion of the Earth Relative to the Luminiferous Ether', *American Journal of Science*, 34, 1887, pp. 333 ff.
Einstein, A., and others, *The Principle of Relativity*, London, 1923, reprinted by Dover, New York.

Michelson Livingston, D., *The Master of Light*, New York, 1973 (a biography by Michelson's daughter).
Swenson, L. S., *The Ethereal Aether: A History of the Michelson–Morley–Miller Aether–Drift Experiments 1880–1930*, Austin, Tex., 1972.
Williams, H. R., *Edward Williams Morley*, Easton, Pa., 1957.
Zahar, E., 'Why did Einstein's Programme supersede Lorentz's?', *British Journal for the Philosophy of Science*, 24, 1973, pp. 95–123, 223–62.

II

Developing the Content of
a Theory

A

Finding the Hidden Mechanism
of a Known Effect

The sciences can be thought of as two-dimensional structures. In the horizontal plane, so to speak, are represented well-established correlations between observed causes and their observable effects. But theoreticians and experimentalists pursue studies in a second dimension, to try to find the mechanisms which produce the observable correlations. The two cases to be described in this section illustrate two different ways of pursuing the quest for explanation. The experiment of **F. Jacob** and **E. Wollman** involved an ingenious isolation and manipulation of the hidden mechanisms of heredity, mechanisms which were later revealed to more or less direct observation by the use of electron microscopes. **J. J. Gibson** demonstrated that a whole new theoretical orientation was required to understand the known human powers of perception, an orientation which had many direct and indirect consequences. In each case the experiment enriched the content of a field beyond the phenomena immediately observable at the time.

12. F. Jacob and E. Wollman

The Direct Transfer of Genetic Material

François Jacob was born in Nancy in 1920. His father was a company director. He was educated at the Lycée Carnot and then at the University of Paris. From 1950 he has worked at the Institut Pasteur in Paris, in a group of molecular biologists of the highest distinction, including Jacques Monod. He became Laboratory Director in 1956. Since 1960 he has been in charge of the whole programme of cellular genetics at the Institut. Though he has become very well known for his experimental researches, carried through with great ingenuity and economy of means, he has not confined his writing to the reporting of his laboratory work. Like many distinguished French intellectuals he has also commented on the general philosophy of his science. His work, *The Logic of Living Systems*, is a discussion of the general theory of biology. In 1965 he was appointed Professor at the Collège de France. In that year he was awarded the Nobel Prize, sharing it with Jacques Monod.

In 1947 he married Lysiane Block. According to an intriguing note in *Who's Who in France*, Professor Jacob admits to taking a greater interest in his hobby, painting, than in his official vocation, microbiology.

Élie Léo Wollman was born in 1917. He was educated in Paris, and has been one of the group at the Institut Pasteur. He became Vice-Director in 1966.

The biology of inheritance

Modern theory of the mechanism by which offspring inherit the characteristics of their parents begins with the discoveries of Gregor Mendel. He is best known for his formulation of the laws of inheritance and the idea that there are dominant and recessive characteristics. This idea suggests that an adult can carry a latent tendency to pass on to its offspring characteristics

138

F. Jacob
and R. Wollman

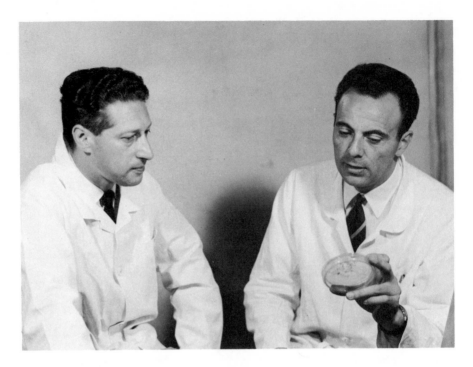

Fig. 79: E. Wollman
(left) and F. Jacob
(right) in their
laboratory at the
Institut Pasteur,
Paris.

which it does not itself develop. The simplest explanation of this phenomenon, and other features of the process of inheritance, is that there are genetic factors, physical 'things' which are present in living creatures, and which are responsible for inheritance. These physical 'things' came to be called 'genes'. Where were they located in the physical body of an organism?

Microscopical studies of the division of cells had shown that there were rod-like bodies in the nuclei of cells that split into pairs during reproduction. One member of each pair migrated to opposite ends of the cell. In this way a pair of nuclei were formed, and the cell divided so that one was in each daughter cell. The rod-like bodies could be made visible by staining the cell with a dye, and so came to be called 'chromosomes', the 'coloured bodies'. For various reasons it began to seem likely that the genetic units, or genes, which controlled heredity, were associated with or perhaps even were the component parts of chromosomes.

Three problems were posed by this theory and its associated images. What was the chemical constitution of the chromosomal material? What were the genetic units or genes, that is how was the structure of a chromosome related to its genetic potential as a carrier of the physical basis of inherited characteristics? How were the genetic units organized, for instance did they lie along the strand of material that had been identified as the chromosome? The first question required a chemical answer; the second a biological solution; and the third required an understanding of how the chemical components of the material of the chromosomes were related to the biological units or genes.

The solution to the first problem, the purely chemical question, came from the work of J. Watson and F. Crick, who found a convincing interpretation for the diffraction photographs of chromosomal material made by M. H. F. Wilkins. Watson and Crick showed that the best explanation for the patterns to be observed in the photographs of the chromosomal material was a double helix, of two strands wound together. Each strand was made up of a sequence of only four chemical constituents. If genes were also physical constituents of chromosomes, then one was driven irresistibly to the conclusion that the biological units must be sets of the basic chemical units. In subsequent work Crick, Brenner and others came to the conclusion that three chemical units made up one biological unit or gene. This has turned out to be no more than a rough approximation, and it seems that the relation between the chemical units and the biological units is rather more complex.

If there is even an approximate relation between genes and chemical units the third question can be answered by mapping the genetic units on to the physical layout of the chromosome. Until very recently no direct and generally applicable method

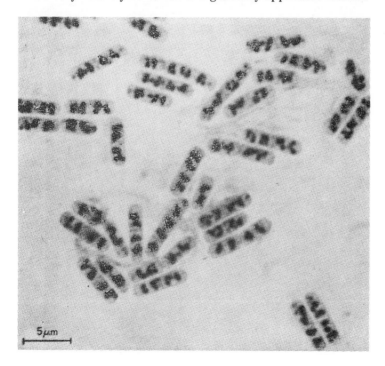

Fig. 80: Microphotograph of chromosomes of the bacterium *Bacillus cereus*, reproduced from W. Hayes, *The Genetics of Bacteria and their Viruses*, 2nd edn, New York (1976), plate 26.

F. Jacob
and R. Wollman

had been found for tracing a given bodily feature back to the place on the strand of genetic material (or DNA as it has come to be called) which was the ultimate source of that feature in the growing organism. All kinds of ingenious indirect methods of 'genetic mapping' had been worked out. The basic idea behind these methods depends on picking out sets of characteristics which are inherited together. In sexual reproduction half the genetic material from the mother cell is recombined with half the material from the father cell to form a complete new complement of genes. In such recombination all sorts of

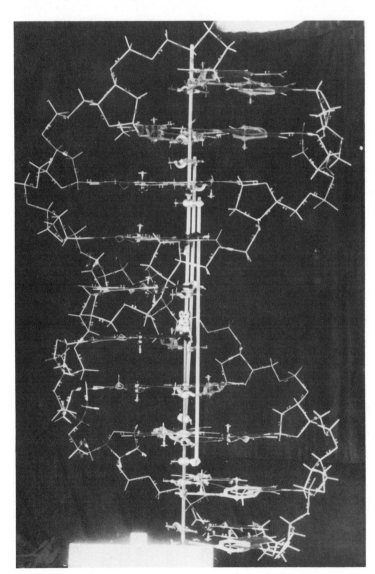

Fig. 81: A model of a fragment of DNA showing the double helix, from J. D. Watson, *The Double Helix*, Atheneum, New York (1968), p. 206.

different associations can be formed. By studying the statistics of recombination in many generations it is possible to make pretty good guesses as to which pairs of characteristics are close together on the genetic material, since they will tend to be inherited together much more frequently than those which are further apart. By working patiently on hundreds of such pairs it is possible to build up a map of the genetic layout of a chromosome.

Two lines of further development suggest themselves. Could some more direct way be found of determining the order

The Direct Transfer
of Genetic Material

Fig. 82: A genetic map of the bacterium *Escherichia coli* strain K12, showing the order of genes. The symbols on the exterior of the ring represent genes, defined in a standard notation expressing their effect on the living organism. From Hayes, op. cit., fig. 127.

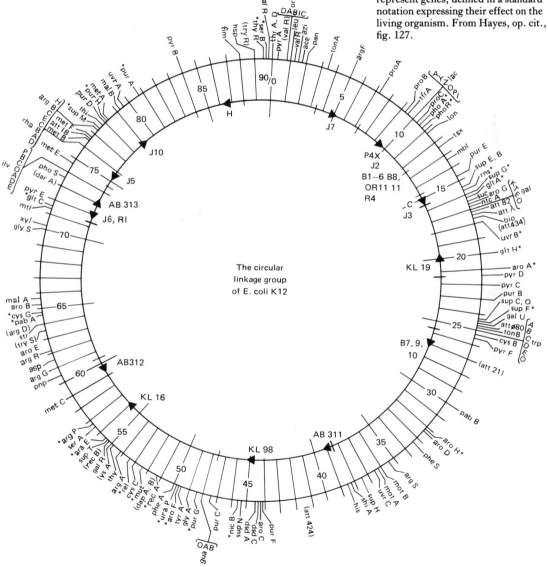

The circular
linkage group
of E. coli K12

F. Jacob
and R. Wollman

of genes and locating sets of them on the chromosomal material? If genes are physically located in this way could methods be developed for excising some genes and replacing them by others, simply by cutting out and putting back lengths of DNA? The experiment to be described in this section answered both these questions in the affirmative, at least in principle. Jacob and Wollman showed how to determine the order of a set of genes more or less directly, and, at the same time, they found that there was a mechanism by which genetic material was lost from and inserted directly into the DNA of single cells.

The experiment

To follow Jacob and Wollman's method for directly transferring genetic material from one cell to another some preliminary information is needed, and some technical terms must be introduced. With some exceptions plants and animals are reproduced cell by cell, through different kinds of process. Cells proliferate by simple division, 'mitosis'. Daughter cells have just the same genetic constitution as the mother cell from which they come. These are called 'vegetative' cells. The other process involves two cells as parents. Half the genetic material of each parental cell migrates to each end, and both cells split in two, to form 'gametes'. Each has half the genetic material needed for reproduction of a complete offspring. One gamete from each parental cell come together to form a pair, and fuse, with each contributing its half to the total complement of genetic material. This process is called 'meiosis'.

There are some cells which can divide only by mitosis. Cells which divide by meiosis must have a double complement of genetic material in their complete state. Cells with two sets of chromosomes are called 'diploid' cells. Those with only one set of chromosomes in the complete form are called 'haploid'. Clearly haploid cells can divide only by mitosis. Most bacteria are haploid, so they must reproduce by mitosis, and in consequence there would seem to be no way in which genetic mixing could take place, as it does naturally in division by meiosis.

If there is no meiosis it seems that it would be impossible for one strain of bacteria to pass on any of their heritable characteristics to another strain. Yet as long ago as the 1920s, it had been shown that dead virulent pneumonia bacteria could somehow pass on their virulence to live non-virulent strains. In 1952 Hayes found that a very small proportion of a common bacterium, *Escherichia coli*, could pass on heritable characteristics independently of cell division and reproduction. In terms of the Watson and Crick discoveries it seemed that some part of the DNA molecule was being passed directly from one bacterial cell to another. Hayes was able to distinguish strains

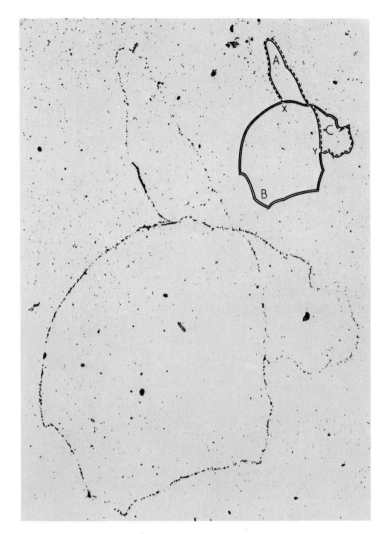

Fig. 83: The splitting of a ring of DNA of the bacterium *Escherichia coli* strain K12 (Hfr♂), starting at X and dividing at Y. The chromosome is about two-thirds replicated, XBY and XAY being the daughter replicas, and XCY being the parental double helix. Reproduced from Hayes, op. cit., plate 27.

of *E. coli* which tended to donate genetic material from those able only to receive it. He called the former F+ and the latter F−. At that time no one had much idea as to what the F+ factor could be. At about the same time Cavalli-Sforza identified an even more active donor strain of the same species of bacterium. He called this the Hfr or High Frequency of Donation strain. When Hfr strains were crossed with F+ strains the result was a strain which showed an even greater tendency to donate genetic material to F− cells.

In many bacteria the genetic material forms a ring. Reproduction starts by the ring forking, at a weak point. The split then runs right around the ring. The ring is made up of a strand of the chemical substance DNA, which is itself a double

F. Jacob
and R. Wollman

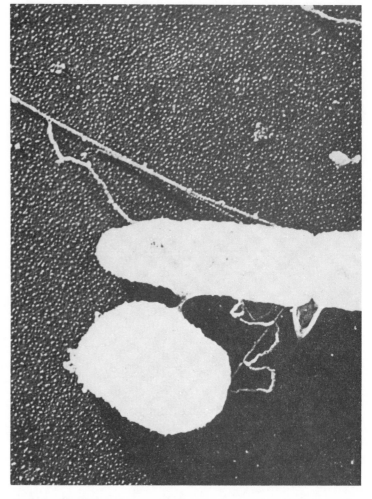

Fig. 84: Electron micrograph of
conjugation of bacteria. The upper
bacterium, undergoing division,
belongs to the *E. coli* K12 donor strain;
the lower is a recipient strain. Hayes,
op. cit., plate 32.

thread wound round itself. It unwinds at the fork at the
extraordinary rate of 10,000 revolutions per minute. Sometimes the point of weakness where the forking starts may break
apart, with an active tip, an 'origin', as it has been called. The
breaking of the ring in this way occurs only in Hfr and F+
cells. It seems to be due to a molecular constituent of the DNA
itself, and so is heritable.

When a break occurs in a cell which is close by an F− cell,
the active origin tends to break out of its original cell and enter
the adjacent cell, pulling with it the remains of the DNA
thread attached to it. The inserted fragments can enter the
DNA of the F− cells, changing its genetic constitution
directly. Jacob and Wollman took advantage of this phenomenon to find the order of genes on the fragment of donor
DNA that was drawn into the recipient cell. Their basic piece

of equipment was a kitchen blender. If the process of insertion
could be stopped, and the strand which is being pulled in by
the F− factor at the active tip of the thread be broken off, a
small number of genes would be slowly drawn in. As each gene
becomes attached to the DNA of the receiving cell it will
change the capacity of that DNA to manufacture proteins and
so alter the behaviour of the whole cell. This suggests that
'there should be a definite relationship between the times at
which a given marker [that is a gene which has a definite and
detectable effect on the behaviour of the recipient cell] is
transferred from the Hfr to the F− cell and the location of the
marker on the Hfr chromosome.'

The experimental technique was very simple. Jacob and
Wollman mixed up a culture of the right combination of strains
of bacteria. They knew from previous experiments that it took
about 2 hours for the process of transfer of genetic material
from Hfr to F− cells to be complete. To stop the process of
transfer with only a short piece of donor DNA drawn into the
recipient they simply switched on the kitchen blender and beat
up the mixture, so physically rupturing the fragile strands of
DNA. By diluting the mixture they prevented any fresh
contacts being made. They tested the resultant culture every
few minutes to see which properties had been passed on from
the Hfr to the F− cells. By 'plating' the cells, that is putting
them on a nutrient jelly, which had been dosed to kill off the
Hfr strain only, they were able to test for Hfr behaviour in the
surviving F− cells.

As Jacob and Wollman put it, 'the extremity, O (for 'origin')
enters first, to be followed by T+L+ [a specific combination
of markers] eight to nine minutes later, and then by other
markers in the order of their arrangement on the chromosome
and at intervals of time proportional to the distance between
them . . . until the whole segment had been transferred.' This
took about 35 minutes. So the order of recovery of the
characteristics associated with the markers gave a perfect map
of the order of the physical layout of the marker genes on the
DNA of the cells from which the fragment had come. By
making use of the fact that on different occasions the markers
are incorporated at different points in the DNA of the Hfr and
F+ strains, leading to breaks at different points in the ring of
DNA of the host cells, a great deal of the structure of the DNA
chain can be explored directly.

By studying the usage of energy by different strains of
bacterial cells Jacob and Wollman were able to find further
indirect proof for the picture of genetic transfer I have
sketched in this section. By starving the Hfr cells but feeding
the F− strain, they were able to stop the process of transfer.
This suggests that only the Hfr cells were using energy in the
initiation of the transfer. But once contact had been made and

F. Jacob
and R. Wollman

the threads broken off it seemed that the F− cells were the source of the power needed to draw in the thread. In a dilute mixture in which little or no contact was being made but in which the process of transfer had been started in the richer mixture, cooling to just above freezing point stopped the drawing in of the fragment of donor DNA, while by warming the mixture to about 37° they were able to restart it.

This experiment introduces another important idea. In some of the earlier examples the discovery of new phenomena, such as magnetic dip, prompted the scientists involved to invent explanations. To explain magnetic effects Norman and Gilbert proposed the hypothesis of a magnetic field. Such a strange kind of thing is very hard to imagine in detail, and it seems unlikely that any experiment could show us the field, in itself. Jacob and Wollman's experiment, however, does seem able to reveal the mechanism by which a strange effect, the transmission of characteristics from one strain of bacteria to another, is brought about. Here we have a more or less direct test of an explanatory hypothesis. Thanks to the electron microscope the delicate threads of DNA can actually be photographed bridging the gap between one haploid bacterium and another. These photograps add a dimension to the actual experiment, which refers to the mechanism of direct genetic transfer only very indirectly and via a network of inferences, the validity of which depends upon our being ready to accept the general picture. Each step in the Jacob and Wollman reasoning is hypothetico-deductive. Suppose that a segment of DNA is being drawn in, what should we expect? It is the expectations that are being tested in the experiment and its subsidiary investigations. Only with the electron microscope photographs is the cycle of reasoning completed, by a more or less direct verification of the hypotheses behind the testable expectations.

Further reading

Wollman, E. L., Jacob, F., and Hayes, W., 'Conjugation and Genetic Recombination in E-coli K-12', *Cold Spring Harbor Symposium on Quantitative Biology*, XXI, 1956, pp. 141–48.

Brown, W. V., *Textbook of Cytogenetics*, St Louis, 1972, ch. 20.

Lewis, K. R., and John, B., *The Organization of Heredity*, London, 1970.

Watson, J. D., *The Double Helix*, London, 1968.

13. J. J. Gibson

The Mechanism of Perception

James Gibson was born in 1904, in McConnelsville, Ohio. His family were strict Presbyterians, and he was brought up in that faith. He did not retain it into adult life. His education was completed at Princeton University, where he took his PhD in 1928. For many years he worked at Smith College. At that time Smith was an institution exclusively for women. He married Eleanor Jack, in 1932. His wife was also a psychologist and much of his subsequent work was done in collaboration with her. Perhaps the famous cookie-cutters came from the Gibson kitchen.

After many years at Smith the Gibsons moved to Cornell University, where, apart from interruptions in the war and several extended visits abroad, James Gibson spent the rest of his life. During the Second World War he worked with the Air Force on the problem of effective training programmes for pilots. He is credited with the discovery that has revolutionized instruction in landing an aircraft, that whatever the angle of descent, the only point ahead which shows no parallax, that is does not change its relation to other things in the environment as the plane descends, is the point at which the aircraft will touch down. It has been said that this was the occasion of his discovery of the important role played by geometrical invariants in the process of human perception.

Fig. 85: J. J. Gibson.

Gibson's work centred round one main problem: how does a human being succeed in perceiving *things*? It was to the systematic exploration of the conditions under which that feat is achieved that he devoted nearly all of his research effort. As with other great experimenters like Michael Faraday, the direction of the programme was determined and its momentum sustained by a profound theoretical idea, that men actively explored a structured world with perceptual systems that had evolved for sustaining life in that world.

In retirement he remained active, taking part in the

beginnings of a new interdisciplinary graduate programme in the neighbouring campus of the State University of New York. He died in December 1979.

Early work on perception

The study of perception, until transformed by the experiments of Gibson, was based upon a pervasive but unexamined assumption. It was supposed that perception was an essentially passive process. The sensitive surfaces of the organs of perception, such as the retina in the eye, were thought to be mere receptors of stimuli. The psychology of perception was not concerned with the relation between things in the world and human experience, but with the relation between the effects of the thing perceived on the sense organs and the thing as perceived. Boring, summing up the trend of research from about 1870, remarks, 'the stimulus has, in general, migrated from the external world to the retina ... the nature of the proximal stimulus at the retina can be predicted and used as the independent variable in experimentation.' It was also supposed that the act or process of perception consisted in the integration of sensory elements into some kind of whole, the thing-as-perceived. In the philosophical version of the basic theory of perception, the doctrine called the sense-datum theory, things as perceived were said to be logical constructions out of sense-data. In traditional epistemology perception was thought of as an ordering of sensations or sensory elements as these made up organized visual, auditory or tactile fields. A tomato-as-perceived was supposed to be a kind of organized sum of red patches of differing shades and tones co-present with tactile, gustatory and other sensations.

The method for experimental study of perception followed from these assumptions. An experimenter should maintain a 'subject' in as passive condition as possible, including physically constraining him, so that he becomes a pure receptor. The presumed components that make up the total perception are added one by one. The results of this programme of research were both equivocal and alarming. A subject held in a rigid frame, and so in a completely passive state, not only did not perceive the world as a world of things, but after a short time stopped perceiving anything at all. This extraordinary discovery ought to have alerted psychologists that there was something radically wrong with the assumptions which had engendered the traditional methods. Psychology is the most conservative of all scientific specialisms, and not surprisingly workers involved in the study of perception continued essentially the same kind of experiment as had been initiated in the late nineteenth century. The breakthrough came in a series of studies which went to the heart of the matter, by querying the very assumptions upon which the methods used in the old work had been based.

Gibson's hypothesis

The old theory could be summed up as follows: 'When the senses are considered as channels of sensation, one is thinking of the passive receptors and the energies that stimulate them . . . it does not explain how animals and men accomplish sense-perception' (J. J. Gibson, *The Senses Considered as Perceptual Systems,* p. 3). The basis of the new theory is a simple but deep observation that while many changes in stimulus energy occur as an organism moves about its environment certain 'higher-order variables' of stimulus energy, that is certain ratios and proportions, do not change. As Gibson noticed, 'these invariants correspond to permanent properties of the environment'. If this is the case, perhaps it is the *change* of sensations engendered by the organism moving about and changing the orientation of its sense organs to some of the fixed features of the physical world that is the major activity needed to produce perception. On this basis Gibson formulated a new theory. 'The active observer gets invariant perceptions despite varying sensations. . .' 'The movements of the eyes, the mouth, and the hands . . . seem to keep changing . . . the input of sensation, just so as to isolate over time the invariants of the level of input at the level of the perceptual system.' In short, we learn or perhaps are programmed to explore our environments, actively seeking the invariants in the relations between changing sensations. These invariants represent the permanent structures of the physical world. We do not, it seems, passively receive and automatically integrate sensory elements into the structured objects of perception.

The great cookie-cutter experiment

Out of a host of ingenious experiments Gibson devised to demonstrate the advantages of the new theory, the most convincing and the most elegant was his demonstration of the role of exploration in the search for invariants in the perception of shape. The experiment required no more elaborate apparatus than most kitchens can provide – a set of metal shapes which are used to cut out dough for the preparation of biscuits (cookies). Until Gibson's pioneer work, psychological experiments always researched the passive rather than the active senses, a tactic based upon Müller's Law – that each excited nerve has a specific conscious quality – and the assumption that perceptions are constructed from sensations, one to each excited nerve.

But when the senses are considered as perceptual systems,

for example when the hand is considered as a *system* consisting of sensitive skin, moving fingers and wrist, with receptors in the joints that register movement, environmental invariants can be detected through continuously changing sensations which stimulate neural *structures* corresponding to invariant structural properties of the things and movements perceived. A very simple experiment that the reader can easily undertake for himself demonstrates this. As one moves one's head from side to side, the world is perceived as stationary, but it is quite clear that the images on the retina at the back of the eye that are cast by the moving lens must be changing in both shape and position. Yet the world is perceived as stationary. But when a moving thing passes the head, and the head is relatively stationary, there might be very similar changes in the images cast on the retina, but now the thing in the world is seen to move, and the head is accepted as the stationary frame of reference. This distinction in perception cannot be based on what is happening to images cast on the retina. The great cookie-cutter experiment carries this thought one stage further and applies it to the field of tactile experience, the way one feels the shapes and textures of things with one's hands.

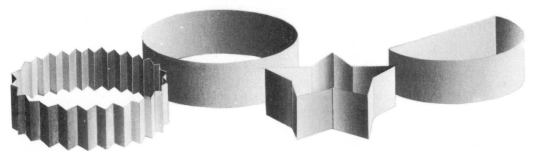

Fig. 86: Cookie-cutters, of the kind used by Gibson in his experiment.

To demonstrate that tactile perception of shape does not come about by adding up patterns of stimulus on the skin, the cookie-cutter experiment was devised. The experiment shows that the perception of shape is the product of the active use of a perceptual organ or system, the hand and its arm. Changing skin stimulation patterns and changing joint orientations help the active human agent as observer and explorer of the world to identify invariants of structure.

Here we have a variety of shapes, each distinguished geometrically from the others by numbers and dimensions of vertices, angles and sides. In the first experiment the shapes were pressed on to the skin of the hand with a standard pressure. This was the passive condition, with the participant's hand held still and without him being able to see what was going on. In this condition the participants could manage correct identifications of the cutters in 29 per cent of the cases. But in the active condition the participant was permitted to explore the shape in any way he liked. He was able to move his fingers actively to bring about a relative motion between the sensitive skin and the object. At the same time the active exploration altered the orientation of wrist and hand and finger joints. In the active condition participants were right in 95 per cent of cases.

This might seem a knock-down demonstration, but a subsidiary experiment was required to eliminate a residual possibility. In the passive condition the hand was held palm upwards and the cookie-cutters were pressed on to the sensitive skin of the palm. In the active condition it was the finger tips which were mainly in contact with the cookie-cutter. Now it could be that the superior results in the active condition came not from active exploration but from the superior capacity of the skin of the finger tips to detect passively experienced patterns of stimulation. To eliminate this possibility it was necessary to bring about relative motion between each cookie-cutter and the palm of the hand, as if the hand were itself actively exploring the shape by running the palm across it. In this condition the whole perceptual system of hand and arm is not fully engaged since the exploratory power of the fingers with their capacity to change shape is not used. The cookie-cutters were once more pressed into the palm of the hand but this time they were also gently rotated. These manipulations produced a sort of pseudo-exploration. In this condition the participants were right as to the shape of the cutter in 72 per cent of cases. The best explanation of this striking result must surely be that it is the changing pattern of sensation that somehow determines the perception of shape. The most economical account of why that might be is just the idea of the perception of invariants. Throughout the changing sensations in the skin the geometrical properties of the shape of the cutter would be the only constants represented through certain invariances in ratios and proportions of sides and angles as the cutters were rotated.

'Tactual perception', says Gibson, 'corresponds well to the form of the object when the stimulus is almost formless, and less well when the stimulus is a stable representation of the form of the object. ... the role of exploratory finger movements in active touch would then be to isolate the invariants ... in the flux of sensation.'

In this simple experiment Gibson demonstrated that active exploration, not passive reception, is the essential process in the way we perceive the things in the physical world.

Later work

Gibson continued his studies to include all the sensory modalities, including that which most interests men, the visual senses. How did Gibson resolve the paradox of the perception of motion, a paradox referred to at the beginning of this section? When the image cast by the lens of eye on the retina moves because of the motion of the observer the world seems to stand still, but when the same retinal motion is produced in a stationary observer by the movement of something in the world, it is the thing in the world that is perceived as in motion. At the back of the paradox, Gibson pointed out, is a philosophical confusion, a muddle about concepts. 'Motion *of* the retinal image is a misconception ... motion *in* the retinal image, change of pattern, is not displacement with reference to the retina.' In perceiving visually the retina is displaced *over* its image, exploring it. For a human being to see, the eye-ball must vibrate some 50 times a second, 'visual tremor' as it is called. This ensures that the fovea, the spot of greatest visual sensitivity, rapidly scans the image, exploring it for higher-order invariants, that is for constant ratios and proportions. By shifting our mode of thought about the relation between image, retina and fovea, the facts, which were well known, fall into an intelligible pattern. We can understand the point of the visual tremor, once we cease to think of the retinal image as moving over the retina, but of the retina running over and exploring the visual image.

Fig. 87: Illustration of geometrical invariants through change of point of view. Adapted from J. J. Gibson, *The Senses Considered as Perceptual Systems*, Boston (1966).

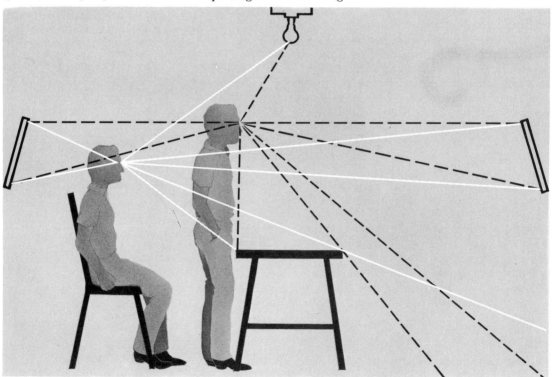

Gibson's pioneering work has led to the definition of a new field of study, ecological optics. This is the investigation of the way the orientation of the body, conceived as an exploratory system, and the visual system centred in the eye co-operate in the active exploration of the energy flow that bathes the human organism, searching that flow for invariants. There are many invariants in the stream of energies, but the human perceptual system seeks only those which have, over hundreds of millions of years, proved valuable in survival of organisms to maturity. These are the invariants which are coordinate with the solid geometry of the sources of much of that energy flow, the material things of the world. Perception is not based on the structure of light as it falls upon the retina, the erroneous theory of the passive, sense-datum theory, but on continuous modifications brought about by retinal movement which co-operates with body posture to reveal its invariants.

One of Gibson's more remarkable discoveries, coming out of his studies of ecological optics, is the apparent paradox that some of the information that is encoded in our perceptions of the world is not given in sensation at all. Information essential for perception flows to the brain from proprioceptors (sensitive nerve-endings) in the joints and muscles. But it cannot be registered consciously as a feeling, or sensation, though it is quite essential to the identification of the structures that we do consciously experience and knowingly explore.

Further reading

Gibson, J. J., 'Observations on Active Touch', *Psychological Review*, 69, 1962, pp. 477–91.
Gibson, J. J., *The Senses Considered as Perceptual Systems*, Boston, 1966.

Boring, E. G., *Sensation and Perception in the History of Experimental Psychology*, New York, 1942.
Gregory, R., *The Intelligent Eye*, London, 1971.

B

Existence Proofs

In general, scientific knowledge comprises a catalogue of the things and substances we believe to exist at some historic moment, and the laws of their behaviour. An important class of experiments is concerned with the testing of putative candidates for inclusion in the current catalogue. In the first example, where we examine **A. L. Lavoisier's** 'discovery' of oxygen, the question of whether there was a material basis of combustion was not at issue. Mayow, Scheele and Priestley had all more or less clearly demonstrated that. The experiment served to locate the substance, oxygen, in the correct and proper category of beings. We could call this kind of experiment 'cap-fitting'. **Humphry Davy**'s decomposition of the alkalis revealed a new kind of substance, prepared for by theory, but not previously isolated. We could call this kind of experiment 'bill-filling'. **J. J. Thomson** succeeded in both aspects of an existence proof – he identified a novel kind of being, the ultimate material unit of electricity, and located it in the appropriate category – a novel category developed for just this purpose.

14. A. L. Lavoisier

The Proof of the Oxygen Hypothesis

Antoine Laurent Lavoisier was born in Paris in 1743, and died there by execution in 1794. His father was a lawyer. His mother died while he was still a child, and he was brought up by an aunt. He was educated at the Collège Mazarin, and took his Baccalaureate in 1763, and Licentiate in 1764. Lavoisier managed to pursue two quite distinct careers with a good deal of success. He entered the Civil Service, or what corresponded to it, as a collector of taxes. But from an early age he also pursued scientific studies with characteristic thoroughness and energy. He was confirmed as a member of the Academy in 1769, after a disputed election, and became a salaried member in 1778. He was equally successful as a tax collector and became a Fermier General, the head of a section of the tax collecting system, in 1780.

The quality of his work was widely recognized in his own time, and he was elected to the Royal Society in 1788. The English connection of his work was very strong since he was effectively extending and rivalling the studies of gas chemistry inaugurated by Cavendish and developed by Priestley. In 1771 he married Marie-Anne Paulze. She was a competent linguist, particularly in English, and assisted him materially by translating his work into English and the works of English authors into French. She survived him, to marry Count Rumford, the extraordinary scientific adventurer, known for his observations on heat while he was superintending the boring out of the barrels of Prussian cannons. Berzelius described the formidable way Baroness Rumford presided over a scientific salon, when he visited Paris long after Lavoisier's death.

After the Revolution of 1789 Lavoisier worked for the new state. He was a member of the commission that planned and managed the introduction of the metric system. But the taint of having been an instrument of the old order, and in particular part of its most iniquitous arm, the tax farmers, hung around

Fig. 88: *Antoine-Laurent Lavoisier and his Wife*, by Jacques-Louis David, 1788. New York, Metropolitan Museum of Art, Purchase, Mr and Mrs Charles Wrightsman Gift, 1977.

him. He was arrested and tried, along with other Fermiers. His trial was famous for the apocryphal remark, 'The state has no need of intellectuals.' Whether this doctrine was really advocated or not he was found guilty of conspiracy and executed in 1794.

The problem before Lavoisier

The phenomenon of combustion had been linked to the study of the composition and nature of the air for at least 100 years before Lavoisier's experiments. The essential step had been taken by John Mayow somewhere about 1673. Mayow (1643–1679) systematically studied the diminution of air caused by combustion and respiration, and explained it by the idea that air was a mixture of distinctive particles, one kind of which were absorbed in combustion. In his *De sal-nitro et spiritu-nitro-aereo*, printed in a collection of works called the *Tractatus quinque medico-physici* published at Oxford in 1674, he concludes, '... the air contains certain particles termed by us ... nitro-aerial which are absolutely indispensable for the production of fire, and that these in the burning of flame are drawn from the air and removed, so that [it] ceases to be fit for supporting fire'. Mayow believed that these particles were also responsible for the elastic force of the air, and hence in their absence the air was more easily compressible, diminishing in volume.

Fig. 89: John Mayow's mouse. An illustration from his *Tractatus quinque medico-physici*, Oxford (1674), tab. 5.

Unfortunately, when Stephen Hales set about repeating and extending these studies about 1724, he picked up the point about elasticity and abandoned the theory of distinctive constituents of gases. He explained the effect of burning a candle in a restricted quantity of air by the hypothesis that the amount the volume diminished 'was equal to the quantity of air whose elasticity was destroyed by the burning candle' (*Vegetable Staticks*, p. 131, Expt CVI). Extending this idea to the effect of respiring air, he explained the flaccid appearance of a bladder from which the air had been breathed many times, and the fact that he could 'plainly perceive that my lungs were much fallen', not as 'owing to the loss and waste of the vivifying spirit of air, but may not unreasonably be also attributed to the loss of a considerable part of the air's elasticity'.

Fig. 90: A combustion experiment conducted by Stephen Hales. *Vegetable Staticks* (1738), plate 16.

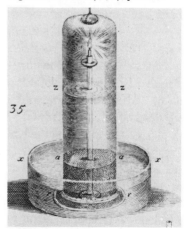

The next step was Priestley's discovery that there were ways of making the respirable part of air by other means. In his *Observations on Air* he describes an experiment of 1774. '... having ... procured a lens of twelve inches diameter and twenty inches focal distance, I proceeded to examine, by the help of it (as a burning glass), what kind of air a variety of substances, natural and factitious, would yield.' By applying great heat to 'mercurius calcinatus per se' [mercuric oxide] he

158

A. L. Lavoisier

obtained an air, not easily soluble in water. 'But what surprised me more than I can well express, was a candle burns in this air with more splendour and heat ...' than in other airs. He tried breathing it, and found it excited an agreeable stimulation. 'Who can tell', he remarks, 'but that, in time, this pure air may become a fashionable article in luxury. Hitherto only two mice and myself have had the privilege of breathing it.'

But, alas, Priestley quite misinterpreted his discovery. He did not follow Mayow's line of thought, that would have led him to identify his 'factitious air' with the respirable part of the atmosphere. He, like Hales, was an adherent of a theory, though a different one, which directed him to interpret these results in quite another way. He believed in the phlogiston hypothesis, the theory that during combustion and respiration a substance (phlogiston) was given out 'which alters and depraves [the air] as to render it altogether unfit for inflammation [and] respiration'. Since he also believed that atmospherical air is a simple, elementary substance, he had to interpret his discovery as the preparation of an air that contained less phlogiston than atmospheric air, so rendering it more fit for supporting combustion and respiration than the atmosphere. 'We can make', he concludes, 'air purer than atmospheric air, that is dephlogisticated air ... containing less phlogiston than the air of the atmosphere.'

Fig. 91: A giant lens commissioned by the Academy of Sciences in Paris, partly under Lavoisier's supervision.

Lavoisier's resolution was simple. He revived, though I think it likely he did not know of it, Mayow's idea that air is a mixture of two 'airs' or gases, one of which is respirable and supports combustion, and the other 'mephitic', unable to support life. In commenting on Priestley's theory and the interpretation of the chemical facts it encouraged, he says, 'My sentiments of the case are different, and I have already given some proof that the residuum of atmospheric air, after combustion, is its mephitic portion, which forms three fourths of its composition, deprived in a greater or less degree of its pure, respirable part.' The experimental test of this view, which would at the same time refute Priestley's ideas, would be to fix the respirable portion by calcination, and then extract it from the substance with which it had combined, and finally restore it to the gaseous residue. If this produced ordinary air again the case would be made. '. . . if, as Dr Priestley supposes, this air [the residual air] were contaminated by some principle which rendered it unsalutary, it would not be sufficient to restore to it the portion of which it had been deprived, but in order to reestablish it in the state of common air, it would be necessary also to separate this contaminating substance from it.'

By a nice irony Lavoisier chose the very same method for the restoration of the lost portion of air that Priestley had used to prepare his pure air, namely the heating of mercuric oxide. But the trick was this: that very mercuric oxide had been itself the product of the slow combustion of mercury in the air sample which was being studied.

The experiment itself Lavoisier describes as follows: 'In a convenient apparatus, which it would be difficult to describe without the aid of engravings, fifty cubic inches of common air were inclosed, to which I introduced four ounces of very pure mercury, which I proceeded to calcine [oxidize] by keeping it, during twelve days, in a degree of heat almost equal to that which it is necessary to make it boil. . . . on the twelfth day, having extinguished the fire and suffered the vessels to cool, I observed that the air which they contained was diminished . . . by about 1/6 of its volume: at the same time a considerable portion of . . . calcined [oxidized] mercury was formed, which I computed to be about forty-five grains. . . . the air, which has thus diminished . . . extinguished candles. . . . In short, it was absolutely reduced to a mephitic state.'

The key piece of reasoning behind the experiment is set out by Lavoisier as follows: 'It has been proved by Dr Priestley's, and by my own, experiments, that calcined mercury is merely a combination of that metal with about 1/12 part of its weight of air, much better and more respirable, if the expression may

The Proof of
the Oxygen Hypothesis

Fig. 92: A lens of the type used by Priestley. This example, from about 1790, was probably made by Joseph Parker of Fleet Street, who made a similar lens for Priestley. It is 16 inches in diameter. Museum of the History of Science, Oxford University.

Fig. 93: Lavoisier's apparatus, illustrated in his *Elements of Chemistry*, Edinburgh (1799), plate iv.

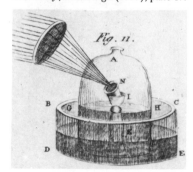

be allowed, than common air: it should appear as proved that, in the preceding experiment, the mercury, as it calcined, had absorbed the best and most respirable part of the air, and left the mephitic or unrespirable part.' Now all that was left to do was to regain the absorbed air and restore it to the mephitic residue. And this is what Lavoisier did. 'I carefully collected the forty-five grains of calcined mercury which had been formed in the preceding experiment; and putting it into a very small glass retort, the neck of which was turned up so as to pass under the edge of a bell glass, filled with and inverted into, water, I proceeded to reduce it without addition. By this operation I recovered nearly the same quantity of air which had been absorbed during the calcination ... when recombined with the air which had been vitiated by that process [it] restored the latter, pretty exactly, to the same state in which it had been, previous to the calcination being performed on it, viz. that of common air.'

It seems, as one might say, that that settles that! There were one or two further steps to make. To bring together oxidation and respiration makes one theory. To this end Lavoisier studied the process of respiration somewhat more closely. He demonstrated that 'the respired air, vitiated by respiration, contains nearly 1/6 of an aeriform acid, perfectly similar to that obtained from chalk', showing that in the process of respiration respirable air was absorbed and, as we should say now, carbon

Fig. 94: General view of Lavoisier's laboratory. Heliogravure from Edouard Grimaux, *Lavoisier, 1734–1794*, Paris (1888). The engraving is based on a drawing by Mme Lavoisier herself, who is depicted taking notes.

dioxide was given out. But some speculations went awry. Here is one: 'These metals form, with highly respirable air, beautiful red calxes ... may we not then suppose that the red colour of blood depends on the combination of dephlogisticated air?' This curious phrase, used instead of the more usual term 'oxygen', reflects Priestley's explanation of his discoveries. If, when oxygen seems to have been driven off from mercury calx by heating and the mercury is restored, then, in the topsy turvy world of phlogiston theory, the result is construed not as a release of oxygen, but as the absorption of phlogiston from the air. The result then must be dephlogisticated air.

The experiment I have described involved little effort at accurate measurement. It simply established the principle of combustion. Lavoisier also attempted quantitative studies of combustion, and given the crudity of the equipment they were of surprising accuracy. By measuring the amount of air that was required to make up for that absorbed during calcination by letting air into a closed retort after it had cooled, he was able to estimate the weight of the air absorbed. In these experiments he used a different metal (tin). By weighing the tin before and after calcination he was able to calculate how much oxygen had been 'fixed' in forming the calx, or oxide. The results were not convincing by modern standards. However, they did show that the loss in weight of the air and the gain in weight of the tin as it turned to calx were compatible with the hypothesis that the same amount of oxygen was lost from the air as the oxide gained.

Further studies of the chemistry of oxygen

The development of the chemistry of gases after Lavoisier involved refinement rather than radical revision of the genre of experiments that had been begun by Mayow, refined by Cavendish and Priestley, and used with such penetration by Lavoisier. Much more accurate quantitative methods were introduced during the nineteenth century, so that strict comparisons between the weights of gases combining in various chemical interactions with the increment in weight of the compounds so formed could be made. Lavoisier himself seriously mistook the place of oxygen in the chemistry of inorganic substances. He seems not to have been able wholly to detach himself from the influence of earlier ideas in which phlogiston, as the matter of fire, was thought to carry a particular activity with it, manifested for instance in the effect it had on the elasticity of the air. Oxygen, he thought, must also be the prime active constituent of acids. It is to that error that we owe the name 'oxygen', 'acid-producer', which he coined for the gas. It was some forty years before the true nature of acids was settled.

How should we interpret the experiment? It would not be quite right to say that Lavoisier *discovered* oxygen. After all both Mayow and Priestley had identified a material. But Mayow discovered *spiritus nitro-aereus* and Priestley discovered dephlogisticated air. It seems that we are not dealing with a factual issue at all, but with testing for the appropriateness of rival conceptions used to marshal the known phenomena into good order. The existence of the 'stuff' is not in question.

Further reading

Lavoisier, A. L., *Essays on the Effects Produced by Various Processes on Atmospheric Air*, transl. T. Henry, Warrington, 1783.

Priestley, J., *Observations on Air*, London, 1774.

Geurlac, H., *Lavoisier – The Crucial Year*, Ithaca, N.Y., 1961.

Leicester, H. M., and Klickstein, H. S., *A Source Book in Chemistry*, New York, 1952, ch. 23.

McKie, D., *Antoine Lavoisier, the Father of Modern Chemistry*, Philadelphia, 1935.

15. Humphry Davy

The Electrolytic Isolation of New Elements

Humphry Davy was born in Penzance, Cornwall, in 1778, the son of a somewhat indigent woodcarver. Davy's father died when he was a child, and his mother, Grace, supported the family by managing a millinery shop. In 1795 Davy was apprenticed to a surgeon. During his apprenticeship he threw himself into a massive project of self-education, including languages and philosophy as well as science.

Evidently this all had good effects, since in 1798 he joined Beddoes's Pneumatic Institute in Bristol, as supervisor of experiments. Beddoes was the centre of a wide circle of literary and scientific acquaintances, and there Davy met both Coleridge and Southey. The former became a very close friend, and was a great influence on Davy, particularly in introducing him to the philosophy of science of Immanuel Kant. A generally Kantian standpoint exerted a great influence on Davy's ways of theorizing. While at the Pneumatic Institute he worked on a systematic study of the medicinal and therapeutic properties of gases. In 1800 he published a book on nitrous oxide (laughing gas). The work was highly successful, and made his reputation. Most of Davy's early scientific writings involved attacks on 'substance' theories of physical action. Typically such theories introduced an unobservable material intermediary to explain the influence of one body on another. For instance, electrical effects were explained by postulating two electrical fluids. He was particularly severe on Lavoisier's use of the mysterious substance 'caloric' to explain the phenomena of heat. True to his Kantian predilections Davy preferred theories based upon the assumptions of attractive and repulsive forces, clearly derived from the theories of R. J. Boscovich, the great Serbian theoretical physicist, and, of course, Kant.

In 1801 Davy was appointed Lecturer at the Royal Institution. He was an enormous success, drawing as many as a thousand people to one of his lecture-demonstrations. He was

Fig. 95: Humphry Davy. Engraving in the Museum of the History of Science, Oxford University.

elected a Fellow of the Royal Society in 1803, and in 1805 was awarded that society's Copley Medal for various applications of chemistry to the practical arts.

From about 1806 he began systematic studies in electrochemistry. He developed the uses of electrical currents as a method of analysis, as I shall describe in the text. Again this was based on a theory of attractive and repulsive forces, and the idea of a physical transport of electricity through the liquids being electrolysed. His theory shadowed forth what we should now call an 'ionic' picture of electrical conduction in solutions. He was convinced that chemical affinity must have an electrical basis. Using his new methods he isolated not only potassium and sodium, but magnesium, calcium, barium, strontium, boron and silicon.

At that time Lavoisier's theory that oxygen was the basis of acids was still widely held. But Davy found that the oxides of his new metals were alkalis. Lavoisier's view of the role of oxygen in acidity must be astray. Part of the explanation, Davy thought, must be that the chemical properties of materials are due not only to the nature of their constituents, but to how these are arranged. By 1810 he had realized that oxygen was not a constituent of all acids. When hydrochloric acid was analysed it yielded hydrogen and another substance, erroneously thought to be an oxygen compound. Since no one, not even Davy, could break it down into constituents, of which

Fig. 96: A popular view of Davy's experiments. This cartoon by Gillray (1802) shows Dr Thomas Young performing an experiment at the Royal Institution. Humphry Davy is assisting.

oxygen might have been one, he concluded that it was indeed an element. And so it has proved to be. We know it as chlorine.

Davy had always been interested in the applications of chemistry and physics to industrial problems, and in 1812 he extended this interest by giving the first courses ever undertaken in chemistry for agriculture.

In 1812 he was knighted and immediately married Jane Apreece, a wealthy widow. She turned out to be a very demanding and tiresome woman, earning a great deal of animosity, not least from Michael Faraday, appointed Davy's assistant in 1813. In that year the Davys and Faraday set out on a continental tour, including a visit to Paris to receive a scientific medal from Napoleon, even though England and France were at war at the time. Lady Davy treated Faraday as a kind of lackey, something which he never forgot or forgave. Returning to England in 1815 Davy was immediately confronted with the task of solving the problem of explosions in mines. From this came his famous Safety Lamp.

There seems little doubt that Davy was, as we should now say, 'into the drug scene'. His contacts among the poets, his own poetic ambitions, and his strongly romantic temperament all conspired to this effect. The visions described in his last but one book, *Consolations in Travel*, have a disconcerting familiarity to those who have read Casteneda and the like. Davy's health deteriorated rapidly in his middle age. After a stroke in 1827 he eked out the rest of his life in isolation and depression, moving from one European resort to another. He died in Geneva in 1829.

Electrolysis before Davy

In order for electrolysis, the decomposition of compound substances by electricity, to be a practical proposition, there had to be a readily available source of steady electric current. The birth of the idea of electrolysis and the development of the technical basis of the accumulator or battery came about together. The first step towards the discovery was Galvani's demonstration, in 1791, that electrical currents can produce muscular contractions. He noticed that a pair of frog's legs, hanging by chance in such a way that they were in contact with a junction between two dissimilar metals, twitched when the metals came into contact. In the years around 1800 Alessandro Volta carried through a systematic study of the excitation of muscular contraction and the production of electricity by the contact of dissimilar metals.

He was able to produce continuous quantities of electricity in a steady flow by assembling a 'pile' of coins, in dissimilar pairs, separated by cards soaked in brine, to provide the contact. The next step was to separate the metals into pairs by arranging them in a series of cups, each containing a conducting solution, and connected by metal strips. This arrangement was called 'the crown of cups'. This was the first wet battery, of which our familiar lead/acid batteries are the direct descendants. The availability of a steady current for long periods of time allowed quite novel effects to be seen. In particular it was soon noticed that gases were produced at the electrodes, and that if the crown included a cup of water, the gases evolved were oxygen and hydrogen.

The experiment

Davy had made various attempts to decompose alkalis by passing electrical currents through aqueous solutions, but he found that 'though there was a high intensity of action, the water of the solutions alone was affected, and hydrogene and oxygene disengaged ...' If the experiment failed, that is potassium metal did not appear when potash was dissolved in

Fig. 97: Volta's 'piles', illustrated in the *Philosophical Transactions of the Royal Society*, vol. 90 (1800).

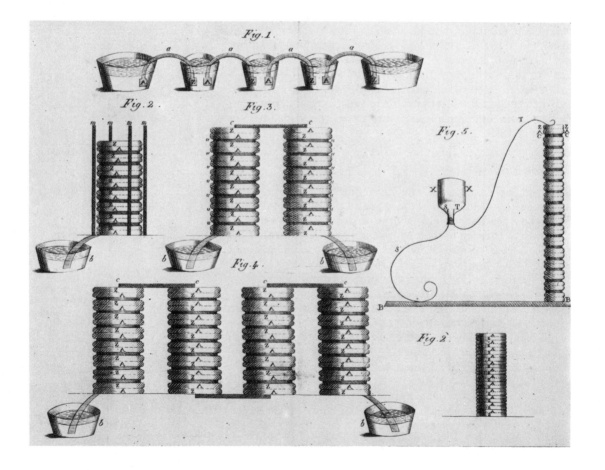

water, what would happen if no water was present at all? So he tried again with molten potash. By heating 'a platinum spoon containing potash, this alkali was kept for some minutes . . . in a state of perfect fluidity'. The effects were spectacular. The spoon was connected to the positive side of the battery and the connection from the negative side was made by a platinum wire which was dipped into the molten potash. There was a bright light at the end of the negative wire and a column of flame rose above the point of contact. But when the polarity was reversed 'aeriform globules, which inflamed in the atmosphere, rose through the potash'.

It was clear to Davy that in these and similar experiments something special was being produced at the negative pole, but it could not be collected and preserved to be closely examined. 'I only attained my object', says Davy, 'by employing electricity as the common agent for fusion and decomposition.' In the experiments with the spoon the potash had to be heated by an external flame. Though solid potash is a non-conductor Davy found that only a little moisture was enough to make it a conductor. In that state it readily fuses and decomposes by strong electrical powers without the uncertainty of the effect of an external source of heat.

Eventually, by this last method, he succeeded. In his biography of Davy, Knight says Davy 'danced round the laboratory' when he finally succeeded in separating the globules. He put a small piece of potash, dampened only by a short exposure to the air, on a round, insulated dish of platinum which was connected to the negative pole of a battery. The positive side was connected to a platinum wire. 'The potash began to fuse at both its points of electrification . . . at the lower or negative surface, there was no liberation of elastic fluid [gas] but small globules, having a high metallic lustre, and being precisely similar in visible characters to quick silver, appeared, some of which burnt with explosive and bright flame, as soon as they were formed, and others remained and were merely tarnished, and finally covered by a white film which formed on their surfaces.'

At last in free form here was the substance he had been looking for, the 'basis' of potash. He soon showed that it was produced independently of the material of which the apparatus was made, so that it must be a constituent of potash. Soda exhibited an analogous result. What were these silvery globules?

When left in the air the metallic globules became covered with a white crust which proved to be potash re-formed again. In pure oxygen the potash crust was formed immediately, but unless water was present to dissolve it, the crust protected the substance underneath from further attacks by oxygen. All the evidence pointed to the simplest interpretation. The experi-

Humphry Davy

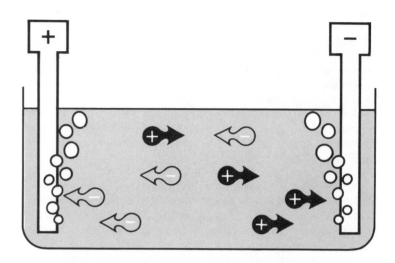

Fig. 98: The migration of ions.

ment had decomposed potash and soda into distinctive 'bases' and oxygen. Davy showed that it was oxygen and only oxygen that was released at the negative pole, while oxygen and nothing else was required to turn the 'basic' back into potash or soda again. This fitted in well with Davy's electrical theory of chemical affinity. As he summed it up, 'The combustible bases of the fixed alkalis seem to be repelled as [are] other combustible substances, by positively charged surfaces, and attracted by negatively electrified surfaces, and the oxygen follows the contrary order.' But when they come together in the synthesis of the potash, 'the natural energies or attractions come in equilibrium with each other'.

Eventually, by patiently collecting the globules (facilitated by his discovery that they did not react with naphtha), Davy made up samples of the new substances. He found that they were good conductors of heat and electricity. But though they resembled metals in all their major properties they were exceptionally light. Their specific gravity was actually less than that of water. They were chemically very active, particularly in reaction with water. He describes several experiments in which these substances seemed almost to be actively seeking out traces of water with which to combine. In a similar way they seemed to hunt oxygen. They would readily reduce, that is extract the oxygen from, other metallic oxides. After consulting the opinions of a number of philosophically minded persons, Davy decided that these 'bases' were indeed metals. He chose to call them 'Potasium' and 'Sodium'. He quickly altered the spelling of the former to our modern 'potassium'. The derivation of these names, he remarks, is 'perhaps more significant than elegant'. But they have 'the great advantage that whether changes occur in the theory of the composition of metals these terms will remain good, for all they mean is the metal derived from potash and from soda respectively'. Davy thought that one should be rather cautious in using terms that were redolent of theory, particularly at a time when discoveries in the electrochemical field were coming so thick and fast.

Caution, too, showed in his observation as to whether these were elementary substances. Probably they were, but all one could say was 'we have no good reason for assuming the compound nature of this class of bodies'.

One further conclusion of consequence could be drawn from the result of this experiment and the testing of the new metals and their chemical properties, and from the study of ammonium hydroxide. There was oxygen in all the alkalis, and in the fixed alkalis, potash and soda, there seemed to be nothing but the metal and oxygen. But Lavoisier had thought that oxygen was the principle of acidity, and indeed that is what the word 'oxygen' had meant. However, says Davy, 'Oxygen then may be considered as existing in, and as forming, an element in all true alkalis; and the principle of acidity of the French nomenclature, might now likewise be called the principle of alkalescence.'

Fig. 99: The great battery at the Royal Institution, illustrated in L. Figuier, *Les Merveilles de la Science*, Paris (1867–70), vol. I, p. 673. This was the main source of electricity for many of Davy's electrical experiments.

Electrolysis after Davy

Further developments of electrolytic methods of decomposition were mostly restricted to industrial applications. The separation of new elements became more and more a matter of chemical analysis. Great analytical sagas, like that of the Curies' separation of radium, were based on finding chemical reactions by which the differential solubilities of corresponding compounds of the elementary substances involved could be exploited to separate them.

If Lavoisier's experiments with oxygen were a case of 'cap-fitting', given the head find the right cap for it, Davy's could be thought of as 'bill-filling', given a prior prescription of what to expect (a light, active metal) how can we find something to fill it? The electrochemical theory had convinced Davy that light metals must be the bases of the oxides. And that theory dictated the means by which they might be released.

Further reading

Davy, H., 'The Bakerian Lecture', *Philosophical Transactions of the Royal Society*, Part I, 1808, pp. 1–44.

Davy, J., *Memoirs of the Life of Sir Humphry Davy, Bart.*, London, 1836.
Hartley, H., *Humphry Davy*, London, 1967.
Knight, D. M., 'Davy' in Gillispie, C. C. (ed.), *Dictionary of Scientific Biography*, vol. 3, New York, 1971.

16. J. J. Thomson

The Discovery of the Electron

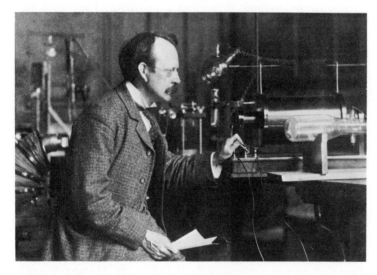

Fig. 100: J. J. Thomson in his laboratory.

Joseph John Thomson was born near Manchester in 1856, the son of a moderately prosperous bookseller and publisher. In keeping with the family's social aspirations he was educated at a private day school. His father hoped that the boy would be trained as an engineer, and had arranged an apprenticeship for him. But while young Thomson waited for a vacancy he began to study at Owen College, at the remarkably early age of fourteen. Shortly afterwards Thomson senior died. No money was available for the apprenticeship premium. Fortunately Thomson won a scholarship to continue his studies at Owen College. Here he came under the influence of the prominent scientists, J. H. Poynting and Sir Arthur Schuster. The latter was an experimentalist of skill, who had made important advances in the study of ionization and the discharge of electricity through gases.

In 1876 Thomson went up to Trinity College, Cambridge, again with a scholarship, to read the mathematical tripos. He had a most successful career as an undergraduate, and was elected a Fellow of his College in 1881. His early work was mathematical, using formal techniques to explore the utility of various mechanical models for electrical phenomena.

He was rather unexpectedly elected as Rayleigh's successor as Cavendish Professor of Experimental Physics, in 1884, though he had little practical experience. In the same year he was elected a Fellow of the Royal Society. He immediately began work on gas discharges, in the context of a long-running controversy between the German and British schools of opinion. The Germans favoured an ether wave theory, and the British a particle picture. This controversy was to lead to the experiments which I will be describing in the text.

Thomson married Rose Paget in 1890. They had two children, a boy and a girl. The boy grew up to be a distinguished physicist in his own right, G. P. Thomson.

During his 'reign' at the Cavendish, 'J. J.', as he was called, not only carried on his own experimental researches, but developed the idea of systematic programmes of research undertaken by groups of scientists. The use of the proceeds of the 1851 Exhibition to finance scholarships and the relaxation of the University's rules about entry of graduate students, permitting people with undergraduate degrees from elsewhere to start research, allowed Thomson to build up a powerful research school. The school not only advanced the subject, but provided trained men to fill chairs throughout the British Empire. The organization of science greatly interested Thomson, and he served on the founding committee of the Department of Scientific and Industrial Research, and as a member of its Advisory Council for eight years. It is said that he had considerable aptitude for finance, and he seems to have made sure that he was personally comfortably off. He was devoutly religious and attended church regularly.

In many ways he came to dominate the British scientific scene as Newton had in his time. He was awarded the Nobel Prize in 1906, and was President of the Royal Society from 1915 till 1920, directing its activities during the First World War. He became Master of Trinity College in 1918. He resigned his chair in favour of Ernest Rutherford in 1919, but continued in active scientific work. He died in 1940.

The atomicity of electricity; the problem before Thomson

The first hints that electricity might be atomic were already present in the results of Faraday's researches into electrolysis, though he did not himself grasp the point. In his statement of the laws of electrolysis, the process by which compound substances were decomposed by electricity, Faraday had clearly formulated two principles. The amount of the product liberated is directly proportional to the amount of electricity passed, and the same amount of electricity liberates masses of products proportional to their equivalent weights. In 1881 Helmholtz pointed out that if one presumes an atomic theory of matter, and accepts the implication that the products of electrolysis are liberated atom by atom, then since Faraday's laws related electricity exactly to matter, there should be 'atoms' of electricity, a definite number for each atom liberated. The proportionality of amount of product liberated to the chemical equivalent rather than to the atomic weight implies that substances whose atoms carry a unit positive charge will require one 'atom' of negative electricity for each atom liberated, those which carry two will require two 'atoms' of negative electricity, and so on.

But this way of interpreting the laws had been missed by Faraday. He held to a field theory of electricity and saw the problem of explaining the laws of electrolysis in terms of structured but continuous fields of force. Most other scientists of the time thought of electricity as a continuous fluid. Without a change in the metaphysical background of physics, the Helmholtz conclusion could not easily be drawn.

The demonstration of the atomicity of electricity turned out to be a long and difficult business. But an essential step in the story was the unravelling of the basis of the phenomena of electrical discharge in rarefied gases. It is to that story that Thomson's great experiment is relevant. We shall see that eventually it connects up most satisfyingly with Helmholtz's way of interpreting Faraday's Laws of Electrolysis.

When electricity is passed through very thin gases sealed in tubes, a great many luminous effects are produced, effects with which we are now familiar in the use of fluorescent tubes for lighting and decoration. It had gradually become clear that some kind of rays were being emitted by the cathode, the metal contact by which the positive pole of the electrical discharge is located in the tube. Davy had shown that an electric arc was deflected by a magnet. J. Plucker took the idea further and systematically tested the power of a magnet to deflect the 'cathode' rays. In his articles of 1858 he seems more interested in the aesthetic than the scientific aspects of his experiments. Shortly afterwards Hittorf demonstrated that the rays cast shadows. If something was put between the cathode and the glass wall of the tube, the glow which the rays produced in the glass was interrupted by the obstacle. The German scientific community had clearly begun to think of cathode rays as a kind of radiation, that is the gas discharges were thought of as ether waves, in the same way as nineteenth-century scientists interpreted light.

J. J. Thomson

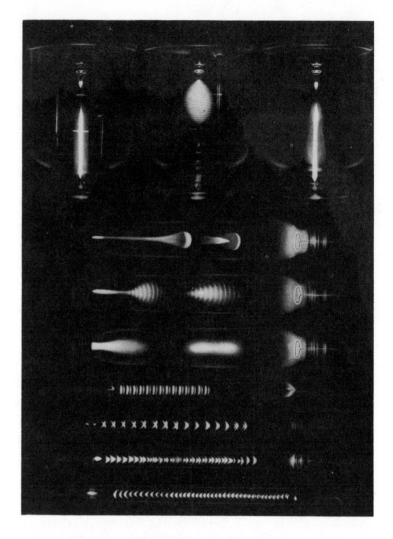

Fig. 101: Plucker's discharge patterns, illustrated in the *Philosophical Transactions of the Royal Society*, vol. 171 (1880), plate 9.

The controversy developed from the attempts to interpret two further experiments. Perrin had proved that the 'rays' had a negative electric charge. He collected 'rays' in a metal tube. When the 'rays' were allowed to enter the tube it became negatively charged, but when he used a magnet to deflect them away from the entrance to the tube, it did not acquire any charge.

But the German group had been doing other experiments. Hertz, in particular, had found that the rays would pass right through thin films of metal, and that these films were not punctured by their passage. How could this fact be reconciled with the idea that they were a discharge of particles? He had also tried to deflect the rays by sending them through an electric field, between two parallel charged plates. But there was no deflection, as there should have been if the particles had been charged. They would have been repelled by the negative plate and attracted by the positive. The rays must be radiant phenomena, that is disturbances in the ether as light was then conceived. If that were so they could not be little 'things' carrying electric charge to Perrin's tube. Instead, they must cause a charge to appear there.

It was to this 'particle versus wave' controversy that Thomson brought his ingenuity for devising and interpreting experiments, and, not least, a determination to carry on a promising line of research in the face of apparently contradictory evidence. His first contribution was made in a report in the *London and Philosophical Magazine* of 1894, that the velocity of the cathode rays was very much less than that of light. This was a severe blow to any radiation theory, since all forms of electromagnetic radiation are propagated with the same velocity.

In a lecture to the Royal Institution of 1897 Thomson set out the ideas that lay behind the experiment to be described in the next section. They consisted of two hypotheses, each presuming that the negative charge detected by Perrin was *carried* by something.

Hypothesis One: 'The size of the carriers [of electric discharge in gases] must be small compared with the dimensions of ordinary atoms or molecules,' because they can penetrate into gases so very much further than comparable beams of molecules, an effect that had been well established by Lenard.

Hypothesis Two: 'The carriers are the same whatever the gas used in the discharge tube.' This had been demonstrated by the fact that the magnetic deflection, using a standard magnet, was the same whatever the gas in the discharge tube.

If these hypotheses are accepted it may be that the 'carriers' are universal, elementary constituents of matter. 'The assumption of a state of matter more finely subdivided than the atoms of an element', says Thomson, 'is a somewhat startling one.'

The experimental proofs that there are electrons

To set about proving his hypotheses Thomson had to show why Hertz had failed to get the expected effect. This was essential if one were to show that the cathode rays were indeed charged particles. It is worth noticing that Thomson did not accept Hertz's result as a disconfirmation or disproof of the particle hypothesis. Rather he used the particle hypothesis to infer that there was something wrong with Hertz's work.

The apparatus involves a cathode as a source of rays, and two metal plugs to act as slits to produce a good beam. Then by fusing wires into the glass it is possible to connect up parallel metal plates D and E to batteries to create an electric field between them. At the rounded end of the tube, a glow appears where the beam of cathode rays strikes the glass.

The first step was to deal with Hertz's failure to get a deflexion in the beam in passing through the electric field. Thomson puts it all very clearly. '... on repeating [Hertz's] experiment I first got the same result, but subsequent experiments showed that the absence of deflexion is due to the conductivity conferred on the rarefied gases by the cathode rays. On measuring this conductivity ... it was found to decrease very rapidly with exhaustion of the gas ... at very high exhaustions there might be a chance of detecting the deflexion of the cathode rays by an electrostatic force.' It was the passage of the rays themselves that prevented their electric charge being detected. By ionizing the gas, as we would now say, the cathode rays effectively short-circuited the plates, and destroyed the electric field.

Fig. 102: The experimental arrangement. C is the source of cathode rays.

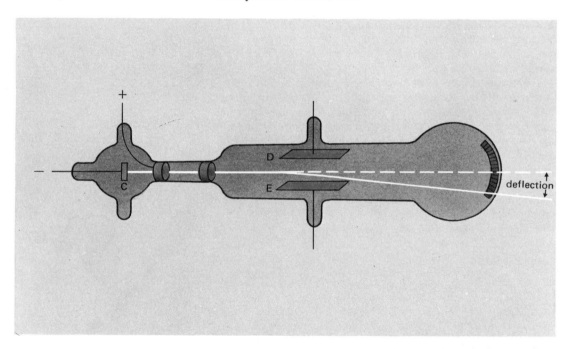

And indeed, Thomson did get just the effect he expected. With a very high degree of attenuation of the gas, using good pumps so that he had a near-vacuum, he found the deflexion. When the plates were connected up so that D was negative and E positive the rays were depressed below the horizontal, but when the plates were connected up the other way round the rays were raised. By pasting a scale on the rounded end

Thomson could measure the deflexion, and he found that 'the deflexion was proportional to the difference in potential between the plates'. Unmistakably there must be an electrical effect between particles and the charged plates. Since he already knew from Perrin's experiment that the rays were associated with a negative charge, he had further confirmation that the rays were carriers of that charge in that the deflexion was away from the negative and towards the positive plate, whichever way round the particular plates were connected.

The heart of the experiment was the measurement of the ratios of the mass of the particles to their electric charge. This ratio was known for other material particles, particularly the charges on fragments of molecules (ions) which are found in solutions. In particular the ratio for the simplest of these bodies, the positively charged hydrogen ion, was well known.

Thomson devised various ways of measuring the ratio of mass to charge (m/e). Readers familiar with contemporary physics will be used to looking for the ratio e/m, but in fact Thomson chose to express the ratio as the reciprocal. The logical extension of his experiment to demonstrate the electrostatic attraction effect was to equip his apparatus with coils to produce a magnetic field. The best way to determine the ratio m/e was to compare the deflexion produced by the magnetic field, call it ϕ, with that produced by the electric field, call it θ. The formula for m/e is $\dfrac{H^2 \theta L}{F \phi^2}$

Now comes a nice Thomson touch. By carefully adjusting the currents producing H, the magnetic force, and F, the electric force, the angles of deflexion could be made equal. This could be tested by so arranging the poles and the plates that the effects cancelled out, and the particles suffered no deviation. Then, by simply switching off the current through the coils the magnetic field could be made to disappear, and thus the deflexion caused by the electric field could be easily measured. If we look back at the above formula for m/e it is clear that if we put $\theta = \phi$ we get the formula $\dfrac{H^2 L}{F \theta}$ (L is the space over which rays are influenced by the fields.) Only one angle needs to be measured, with a corresponding increase in accuracy. The arrangement can be seen clearly in the figure.

Details matter in experiments, and one small detail had to be seen to in this one. The glow on the rounded end of the tube was too faint to be seen easily in daylight. 'As it was necessary', says Thomson, 'to darken the room to see the phosphorescent patch, a needle coated with luminous paint was placed so that by a screw it could be moved up and down the scale. Thus, when the light was admitted the deflexion of the phosphorescent patch could be measured.'

All kinds of variations were made in the materials used in the

J. J. Thomson

apparatus to make sure that the effect was due to something present in or evocable from any kind of material, that is a universal constituent of matter.

The results were as Thomson expected. The value of m/e turned out to be independent of the nature of the gas. And in keeping with the hypothesis that the cathode rays are streams of particles of subatomic dimensions the value of m/e turned out to be very small. In particular it was very small compared with the known value of the ratio for the hydrogen ion in electrolysis. In fact, if the value of m/e for the cathode rays is 10^{-7} units, that for the hydrogen ion is 10^{-4}. In short, if cathode ray particles and hydrogen ions have the same unit charge, e, cathode ray particles are 1,000 times smaller than hydrogen ions.

But as Thomson notes, 'the smallness of m/e may be due to the smallness of m, or to the largeness of e, or to a combination of these two.' There was an argument for smallness of size. Lenard had demonstrated the great penetrating power of cathode rays which suggested that they were smaller than molecules. In the event Thomson found arguments both for the smallness of m and the largeness of e. The latter was wrong. Since each cathode ray particle, it has turned out, electrically balances one hydrogen ion, their charges must be equal and opposite. So the correct conclusion to draw from the comparison of the ratios m/e for each is that the cathode ray is only 1,000th of the mass of the hydrogen ion.

Fig. 103: Thomson's apparatus, now in the Science Museum, London.

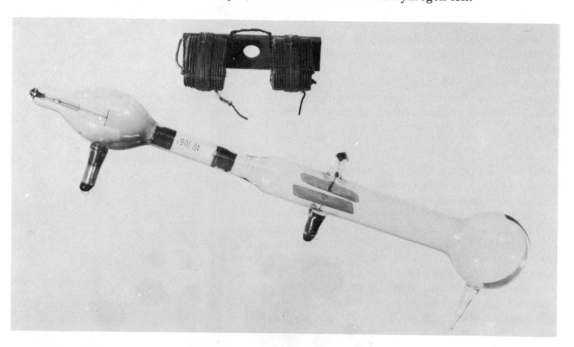

But that was a minor blemish soon put right. The deep speculative conclusion Thomson drew from the experiment has determined the direction of physics ever since. Why is m/e independent of the kind of gas in the tube? Thomson's answer was to develop his sub-atom hypothesis. '... in the atoms of the different chemical elements are aggregations of atoms ... of some unknown primordial substance X ...' In the strong electric fields near the cathode the molecules of the gas are broken up, ionized as we should say, and release a few of their 'primordial atoms'. This conclusion can be confirmed from Lenard's result. He had found that the depths the primordial atoms could penetrate into a substance depended upon nothing but the density of the medium. If the molecule is a spatial arrangement of the subatomic corpuscles, collisions will be between corpuscle and corpuscle, not corpuscle and molecule. Thus collisions will be proportional to the number of corpuscles, primordial atoms, not to the number of molecules. But the number of corpuscles will be proportional to the ratio of the total mass to the volume, that is, to the density of the gas, and hence the mean free path (a measure of depth of penetration) should be inversely proportional to this, since the fewer corpuscles the further a corpuscle should go before collision.

Having made this step it is easy to turn to a theory of atomic architecture. 'This negative ion [his old "primordial atom" or "corpuscle"] must', he says, 'be a quantity of fundamental importance in any theory of electrical action ...' 'When they are assembled in a neutral atom the negative effect is balanced by something which causes the space through which the corpuscles are spread to act as if it had a charge of positive electricity equal in amount to the sum of the negative charge of the corpuscles.' 'A positively electrified atom is an atom which has lost some of its "free mass", and this free mass is to be found along with the corresponding negative charge.'

Finally he concluded that the chemical properties of atoms might be due to the structured organization of these 'corpuscles' in the chemical atoms. The familiar periodic table of the elements, with its families of chemically similar substances, would then be a consequence of the repetition of structures as layer after layer of subatomic constituents was built up.

The 'primordial corpuscle' after Thomson

The general tenor of Thomson's speculations turned out to be right. But his views had to be corrected on matters of detail. The idea that the positive charge was spread through the region of space in which electrons were embedded soon gave place to Rutherford's nucleated atom with the positive charge concentrated in a central heavy nucleus and the electrons as planetary charges orbiting it. At the time of Thomson's major discoveries there was no idea of the quantization of energy, and many other physical parameters soon seemed to be called for to describe the internal architecture of atoms. There only remained the right thing to call these 'primordial atoms of material X'. Following a suggestion of G. H. Stoney, they were soon universally referred to as *electrons*.

Thomson was not just correlating phenomena, but actively seeking effects which would differentiate one 'picture' of the structure of matter from another, a corpuscularian or atomistic 'picture' from a world conceived in terms of ethereal waves.

Further reading

Plucker, J., 'On the Action of the Magnet upon the Electrical Discharge in Rarefied Gases', *The London, Edinburgh and Dublin Philosophical Magazine*, 4th series, vol. 16, 1858, pp. 119–35.

Thomson, J. J., 'Cathode Rays', *The London, Edinburgh and Dublin Philosophical Magazine*, 5th series, vol. 44, 1897, pp. 293–316.

Thomson, J. J., *Conduction of Electricity through Gases*, Cambridge, 1903.

Thomson, G. P., *J. J. Thomson and the Cavendish Laboratory of his Day*, New York, 1965.

Whittaker, E., *A History of Theories of the Aether and Electricity*, vol. I, London, 1951, pp. 348–66.

C

The Decomposition of an
Apparently Simple Phenomenon

Bacon recommended that scientists should study 'the forms of simple natures'. By this he meant the most fundamental knowledge would be of the structural properties of matter responsible for the basic phenomena into which the world of experience could be analysed. At the beginning of the scientific investigation of any field it is vital to demonstrate experimentally which things are compound and which simple, relative to one's method of analysis. Sometimes the experimental work necessary to achieve this is difficult and its results controversial. **Isaac Newton** believed that he had given a final demonstration (*experimentum crucis*) that sunlight, though apparently homogeneous, was a mixture of rays of 'different refrangibility'. He thought that any other way of construing ordinary light was ruled out by the experiment to be described in this section.

17. Isaac Newton

The Nature of Colours

Fig. 104: *Isaac Newton*, oil painting attributed to Sir Godfrey Kneller (1718). Trinity College, Cambridge.

Isaac Newton was born at Woolsthorpe in Lincolnshire on Christmas Day, 1642. His father had died before he was born, and his mother married again when he was only two. As a child he demonstrated his manual dexterity as he 'busied himself making models of wood in many kinds'. Most of his childhood was spent with his grandmother. He went away to school at Grantham, and then on to Cambridge in 1661, but not before he had tried his hand at farming without a great deal of enthusiasm.

Newton was very successful at Cambridge. He was elected to a minor Fellowship at Trinity College in 1667 and became a major Fellow in 1668. In 1669, at the age of twenty-six, he was elected to the Lucasian Chair of mathematics.

The Great Plague had closed the university in 1665, and Newton retired to his mother's farm at Woolsthorpe. His great productive period had begun in about 1664. The falling apple that sparked off his theory of universal gravitation is said to have come from one of the trees in the Woolsthorpe orchard. Between 1665 and 1667 he developed the method of fluxions (the calculus, as we now call it), carried out most of his experimental work on the nature and properties of light, and laid the foundations of the universal mechanics in which he synthesized the terrestrial science of Galileo with the planetary theory of Kepler. But he took many years to prepare these discoveries and inventions for publication. Newton was very sensitive to criticism, and the equivocal reception of his first communication to the Royal Society, on the nature of light, made him wary of publishing mere fragments of research. So we find him holding on to his discoveries until they could be worked up into massive treatises. The *Principia*, the great work in which he set out his mechanics and cosmology, did not appear until 1687. The *Opticks*, most of the experimental work for which had been done around 1666, was finally published only in 1704.

In 1689 Newton took his seat in the House of Commons as a Member for Cambridge. This event marked a considerable change in his interests, and some historians have suggested, in his character. He virtually abandoned scientific research from about this time, and enjoyed the life of a senior administrator and public figure. He became Master of Royal Mint and is said to have run it with exemplary efficiency. Throughout his life he had taken an intense interest in theological matters. Even in old age he was still trying to solve chronological problems in the dating of events recorded in the Old Testament. He died in 1727, having acquired a reputation in his own life-time that no other scientist was ever quite to have again.

Early work on light and colour

Is colour a quality of light produced *in* a body, or is it a quality separated out of light *by* a body? This seems a question of some profundity and its solution likely to be of great technical difficulty. The problem had a long history. Theodoric of Freibourg, whose masterly solution of the difficulties of understanding the rainbow we have studied above, was typical of medieval thinkers in generalizing a vaguely Aristotelian explanation. He thought that light acquired its colour from the medium through which it passed. His explanation is based upon the idea of pairs of contrary principles. A medium can be more or less translucent. Near the surface a medium is more bounded than it is in its depths. A mirror is perfectly bounded, and reflects all light, having no effect on colour. A transparent solid is unbounded, allowing light to penetrate deep into its interior. White light is passed by a medium having a perfect balance of the four contraries. When a medium is relatively bounded, that is near its surface, light is qualitatively changed so as to appear red. But when the medium is relatively opaque in its interior, the light is so changed as to appear blue.

This explanation could hardly be counted very satisfactory since the contraries seemed rather more mysterious than the production of colours they were called upon to explain. A closer study of the way light was affected by transparent objects showed that the colours had something to do with the way light was refracted when passing from one medium such as glass to another, such as air. Descartes was the first to separate light of pure colour using this effect. In *Les Météores* of 1637 he describes an experiment which he had performed in the course of studying the rainbow. The experimental arrangement is shown in the figure on p. 100 above. 'When I covered one of these surfaces with a screen,' says Descartes, 'in which there was a small opening DE, I observed that the rays which pass through this opening and are received on a white cloth or sheet of paper, show all the colours of the rainbow; and that the red always appears at F and the blue or violet at H.'

What relation did these coloured rays have to the light from the sun which had fallen on the prism? It was to the answer to this question that Newton's experiment was addressed.

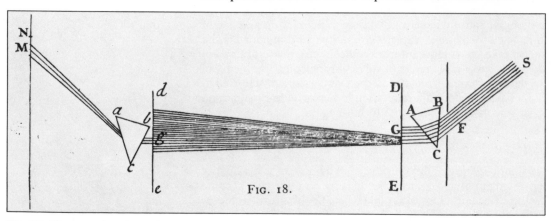

FIG. 18.

Fig. 105: The separation of rays of different 'refrangibility'. Newton, *Opticks* (1721 edn), book I, part I, table iv, fig 18. S is the source of white light. In prism ABC the rays of different refrangibility are separated. The screens DE and de serve to separate progressively purer colours.

Newton's systematic research programme

Newton's series of more and more successful versions of the basic experiment to be described here was not original in conception, but it was to develop into a fairly exact execution. (For an account of the forerunners of Newton in the study of colour and refraction see J. A. Lohne, *Notes and Records of the Royal Society of London*, 20, 1965, pp. 125–39.) In his letter to the Royal Society of 1672, Newton tells of the puzzlement he felt, when in an experiment of 1666, he noticed that the shape of the spectrum image cast on a screen by passing light from a round hole through a prism, was oblong, 'with straight sides' as he says. Why should this be so? According to Lohne (see Further Reading), Newton must have tidied up his description of this image somewhat, since the greater intensity of the yellow component in the sun's light would have made the image rather broader at that point in the spectrum.

In preparing a definitive account of the experiment for the *Opticks*, Newton describes how he took pains to refine and sharpen the image. 'By using a larger or smaller hole in the window-shut [he] made the circular images larger or smaller at pleasure. The amount of light could be increased by using a narrow oblong hole rather than a circular one, keeping the ends of the spectrum image sharp.' Newton seems to have ignored or overlooked diffraction effects of the use of a small hole as image, though these had been noticed by his contemporaries.

The basic experiment, refined by the use of a lens to focus the image of the hole, was quite simple. The spectrum is thrown on a piece of black paper in which there is a small hole. When the hole coincides with the red part of the spectrum a beam of red light is obtained, which can be refracted through a second prism. Similarly when the hole coincides with the blue part of the spectrum a blue beam is separated out. It is the effect of the second prism that is the key. There are two results to be noticed. The resulting image, whatever its colour, is quite circular, 'which shows that the light is refracted without any dilatation of rays', since the shape of the hole is perfectly reproduced in the image. But when a blue ray passes through the second prism it is more refracted than a red ray. So the separation of the colours is a secondary effect. The underlying process is the separation of 'rays of different refrangibility'. In a letter to Lucas of 5 March 1677/8, Newton was at pains to emphasize the true result of the experiment. '. . . you think I brought it to prove that rays of different colours are differently refrangible: whereas I bring it to prove (without respect to colour) that light consist of rays differently refrangible. What the colours of the rays differently refrangible are . . . belongs to after enquiry . . .' (quoted by Lohne).

What is probably the last of Newton's many versions of the experiment is illustrated in the engraving to be found in the Paris edition of the *Opticks*. It was drawn from a sketch supplied by Newton himself (cf. Lohne, 1968).

So far Newton had achieved no more than a more exact repetition of the cruder experiments of his predecessors. Even the testing of monochromatic light by passing it through a second prism had been anticipated, albeit crudely, by J. M. Marci of Kronland. Marci was a prominent physician in Prague. Though isolated from contacts with Western scientists by the Catholic reaction in Bohemia in the early seventeenth century, he did important work in astronomy, optics and medicine. But though he succeeded in decomposing white

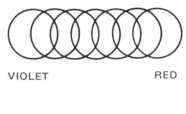

VIOLET   RED

VIOLET   RED

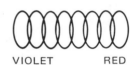

VIOLET   RED

Fig. 106: The effect of using light sources of different shapes.

Fig. 107: An engraving showing Newton's apparatus for separating rays of pure colour. W. J. 's Gravenande, *Physices Elementa Mathematica Experimentis Confirmata*, Leyden (1742), table cxviii.

light into coloured beams, it was to be left to Newton successfully to reconstitute the original beam.

But to demonstrate that the phenomenon of colours in refracted light is caused by the different refrangibility of rays already present in the white beam, and not by some modification produced in the light by the glass of the optical apparatus, something more is needed. Newton's original recombination experiment reported in the Letter of 1672 involved the use of a lens to bring about the confluence of the rays. The reactions of many of Newton's contemporaries to the experiment were tepid. Hooke objected that the experiment does not show that the light, prior to refraction, should be thought of as a collection of these different rays. They could have been produced in the process of refraction. However, in the *Opticks* Newton added another and very ingenious recombination experiment to refute this kind of objection.

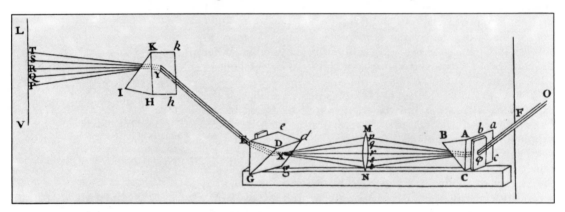

Fig. 108: Decomposition, recomposition and decomposition of white light to the spectrum. Newton, *Opticks* (1721 edn), book I, part II, table iv, fig. 16. Rays refracted by prism ABC are recombined optically by lens MN, and are reseparated by prism KIH.

By using a long, flat prism, Newton made the angle which separates the beams of coloured light very small. By altering the angle of a screen arranged as in the figure on p. 187, colours can be produced from what looks like white light. When the screen is at position B, there is enough diffusion of light caused by dust particles in the air for the narrowly separated coloured beams to be mixed again. By altering the angle of the screen to position C the coloured beams are made to strike the screen at sufficiently separated places for a spectrum to be seen. The distance WZ, separating the points of contact of the red and blue beams with the screen in position C, is much greater than the distance XY separating the images from the red and blue beams when the screen is in position B. The only feature of the arrangement which varies is the angle of the screen. The separation of images is being brought about by manipulating something quite independent of the prism which is producing the original, narrowly differentiated beams. Altering the angle of the screen allows the differently coloured rays to be identified without the diffusion of light from one beam to another which occurs when the images are very close together.

To clinch the matter Newton undertook a much greater variety of optical manipulations than Marci had attempted. Newton showed that once the colours had been properly separated they were unaffected by any of his manipulations. Refraction and repeated refraction did not change the colour.

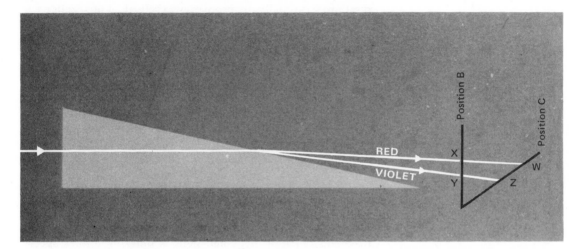

Fig. 109: Recombining colours without a lens.

In a typical refraction experiment Newton illuminated an object with monochromatic light, and then looked at it through a prism. If the passage of light from the object to the eye through the prism had had any effect on the light then he should have seen some difference in the colour of the thing when so observed. 'But those illuminated with homogeneous light appeared neither less distinct, nor otherwise coloured, than when viewed with the naked eye.' Newton remarked that since the differences between the rays might really be continuous, light could not be perfectly homogeneous, no matter how sharply focused. But the spread of colours in each apparently homogeneous ray is so small that 'change was not sensible, and therefore in experiments where sense is the judge, the change ought not to be considered at all'. Truly homogeneous light cannot be produced by refraction. Modern lasers which do produce perfectly coherent light depend upon a different physical principle.

The final step was to examine a wide variety of substances, 'paper, ashes, red lead, gold, silver, copper, grass, blue flowers, violets, bubbles of water tinged with various colours, peacock's feathers and such like . . .' Under red light, they all appeared red. Under blue light they all looked blue, under green light, green and so on. Reflection, like refraction, has no effect on the colour of relatively homogeneous light.

Isaac Newton

But why are these results so readily and unambiguously obtained? Newton and Descartes before him had supposed that in some way or another the motion of particles was involved in the transmission of light. Newton considered the speed of the particles to be the cause of our experiences of colour, while Descartes thought it had to do with their rate of rotation. Eventually the problem was solved, at least relative to the known phenomena, by Euler. About the year 1746 he gave precise mathematical form to another rival theory that had been proposed, notably by the Dutch physicist, Huyghens. Euler showed that Newton's experimental results and many other phenomena could be elegantly explained by assuming that light was propagated as a wave in an all-pervasive medium, the luminiferous ether. Light was not to be thought of as a stream of particles, but as vibration in an elastic solid. Colours corresponded to waves of different wavelength. This explained why different colours were differentially refracted when they passed from one medium to another. The colours were not produced in the medium, as medieval physicists had thought, but at the boundary between media. Elegant though Euler's solutions were, they too have to be modified under the pressure of still more recondite discoveries about electromagnetic radiation of which light is only one rather special kind.

In most of the experiments preceding Newton's study of colour, the subject under investigation lay ready to hand in the common experience of mankind. Falling bodies, compressed gases, the rainbow and its accompanying drops of rain, even the developing chick, are all with the range of our senses. In the conclusion Gilbert drew from Norman's experiment a more subtle kind of being is proposed, something no human observer could ever experience. The *orbis virtutis* is the unobserved or 'occult' cause of observable magnetic effects. For all their apparent simplicity Newton's experiments on colour also go beyond experience, though not so deeply as those of Norman and Gilbert. Newton's refractions and screenings show that white light (which can never be perceived by us as other than white) is 'really' a mixture of coloured rays, which can be perceived as they are, only when separated from all others by some accidental or human manipulation.

Further reading

Descartes, R., *Les Météores*, Discours VIII of *Discours de la Méthode et les Essais*, Leyden, 1637.
Newton, I., 'A letter of Mr. Isaac Newton, ... containing his New Theory of Light and Color' (1672), facsimile reproduc-

tion in Cohen, I. B., and Schofield, R. E., *Isaac Newton's Papers and Letters in Natural Philosophy*, 2nd edn., Cambridge, Mass., and London, 1978, pp. 47–59.

Newton, I., *Opticks*, first published in English in 1704. Reprinted by Dover Books, New York, 1952.

Young, T., 'On the theory of light and colours', *Philosophical Transactions of the Royal Society*, vol. 92, 1802, pp. 20–71.

Lohne, J. A., 'Experimentum Crucis', *Notes and Records of the Royal Society of London*, vol. 23, London, 1968, pp. 169 ff.

Manuel, F. E., *A Portrait of Isaac Newton*, Cambridge, Mass., 1968.

Sabra, A. I., *Theories of Light from Descartes to Newton*, London, 1967.

D

The Demonstration
of Underlying Unity
within Apparent Variety

Complementary to the kind of project described in the last section, an experimentalist might set about trying to show that some apparently diverse collection of vaguely similar phenomena had a strict underlying unity. Perhaps they were each a manifestation of fundamentally the same kind of state or condition of nature. In this section I describe how **Michael Faraday** laid down criteria for 'underlying sameness' and then set about demonstrating, within the margins of precision allowed by experimental technique, that the apparently diverse kinds of electricities were manifestations of a common underlying 'something'. What that 'something' was he did not himself establish.

18. Michael Faraday

The Identity of All Forms of Electricity

Michael Faraday was born on 22 September 1791. His father, James, was a country blacksmith, who had moved to London, just before Michael's birth. Though the family were poor they seemed to have been remarkably close and contented. They were Sandemanians, and the strength of their family life must have had something to do with the intensity of their religious convictions. Michael Faraday was a Sandemanian preacher all his life. The sect had originated in Scotland. Sandemanians hoped to bring about the separation of Church and State, and to reconstitute all the early Christian forms of worship, including the 'love' feast, a substantial communal meal. God was thought to be an active being, working in the world. They favoured a naturalistic proof of His existence through the contemplation of nature.

Fig. 110: Michael Faraday. Photograph in the Museum of the History of Science, Oxford University.

At the age of fourteen Michael Faraday was apprenticed to a bookbinder. In working at this trade he acquired both manual dexterity and a passion for knowledge, since he read the books he was set to bind. In his spare time he attended courses at the City Philosophical Society. He came to the notice of Humphry Davy, serving the injured Davy as an amanuensis. Shortly afterwards he was taken on as an assistant by Davy, at the Royal Institution. Davy and his wife travelled extensively on the continent. Faraday went with them, officially as Davy's scientific assistant, but to his great resentment was treated more like a valet by Lady Davy.

In 1825 he was elected Director of the Laboratory at the Royal Institution. His astonishing capacity for sustained experimental work was coupled with a powerful vision of the basic workings of nature. From Davy he had absorbed the idea of the world as a structured whole, formed by continuously interacting natural agents or powers. With this theory to guide his studies he was soon in the forefront of the sciences of chemistry and physics.

His family life seems to have been very similar to that of his parents. He married another member of the Sandemanian congregation, Sarah Bernard, in 1821. Faraday seems to have been rather a jolly man amongst his intimates. He was devoted to strenuous physical exercise, walking great distances, and he was one of the first bicyclists. But like many of the most active thinkers of that time, he suffered a severe mental breakdown in mid-life. He never fully recovered. His memory began to fail and he had to have recourse to all kinds of devices to keep track of events even in the course of a single morning. In 1858 Queen Victoria provided him with a home near Hampton Court, to which he retired in 1862. His last years were spent quietly, since his capacity for active scientific work had completely gone. He died in 1867.

The problem of the identity of electricities: qualitative preliminaries

In his *Experimental Researches*, Series III, paragraphs 265 to 378, Faraday describes the masterly series of experiments he undertook to determine whether the superficially distinct forms of electricity were merely different manifestations of a common underlying unity. In a sense Faraday already knew, before he undertook the very first experiment, that there was really only one basic electricity. His metaphysics of nature allowed him no other conclusion. But metaphysical conviction is worthless without empirical demonstration.

Why had anyone supposed that there were many electricities? The argument depended on the presumption that if superficially similar effects were produced by quite different processes they must really be caused by quite different underlying entities, whatever those might be. As Faraday notes in 1833, 'many philosophers are still drawing distinctions between the electricities of different sources; or at least doubting whether their identity is proved.' The varieties in question are: 'common', that produced by friction; 'voltaic', that produced by chemical action; 'magneto', that produced in electromagnetic generators; 'thermo', that produced by heating the point of contact of two dissimilar metals; 'animal', that produced by, for example, electric eels.

The experimental programme depends on the argument that if all these electricities can be shown to be identical in their effects, then despite differences in their origins, they must be essentially the same. The key effects are the evolution of heat, the production of magnetism, chemical decomposition, certain physiological effects, and the capacity to produce a spark. The series of experiments is organized as a systematic demonstration that electricities from each of the sources produce all the qualitative effects.

Voltaic electricity is known to be capable of causing most of the required effects, but it remains to establish in a general way that it can flow as a current. If it can be shown to be capable of flowing, then all the typical effects of electrical motion can be expected.

The experimental series

To prove that voltaic electricity can take the form of a current Faraday could have used a galvanometer, an instrument for detecting an electric current. But much more sensitive devices can be constructed. Electric currents will decompose compound substances which have been dissolved in water, even if the current is very weak. By choosing a compound, one of whose constituents becomes visible when released, even if only a little is freed, Faraday devised a very sensitive detector of electric currents.

Fig. 111: Faraday at work in his laboratory. Engraving in the Museum of the History of Science, Oxford University.

He had demonstrated in some earlier studies that currents can flow across air-filled gaps in electrical circuits, when the air is heated. The apparatus to test whether voltaic electricity could flow as a current consisted essentially of a battery, as a source of voltaic electricity, connected to a circuit which included an air-filled gap. When the air in the gap is heated a current should pass immediately, if voltaic electricity could indeed produce one. Here is how Faraday describes the experiment: 'As heated air discharges common electricity with far greater facility than points, I hoped that voltaic electricity might in this way also be discharged. An apparatus was therefore constructed in which AB is an insulated glass rod upon which two copper wires, C, D are fixed firmly; to these wires are soldered two pieces of fine platina wire, the ends of which are brought very close to each other at e, but without touching; the copper wire C was connected with the positive pole of a voltaic battery, and the wire D with a decomposing apparatus from which the communication was completed to the negative pole of the battery. In these experiments only two troughs, or twenty pairs of plates, were used.

'Whilst in the state described no decomposition took place at

Fig. 112: The apparatus for demonstrating a discharge of voltaic electricity.

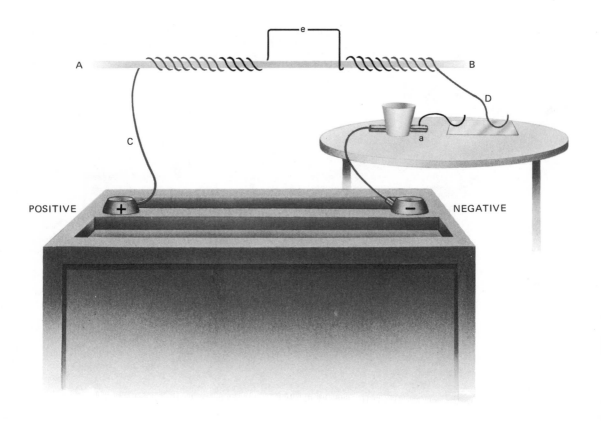

the point a, but when the side of a spirit lamp flame was applied to the two platina extremities at e, so as to make them bright red-hot, decomposition occurred; iodine soon appeared at the point a, and the transference of electricity through the heated air was established.' It was well known that voltaic electricity would produce all the other effects of Faraday's list. So the first step had been taken.

Turning now to common electricity, that produced in one of the friction machines of the day, the same set of effects must be demonstrated. Again many of the effects of common electricity were already known, for instance the heating effect of electricity generated by such a machine. It remained to demonstrate that common electricity has a magnetic effect comparable with that of voltaic electricity.

'If common electricity is identical with [voltaic],' says Faraday, 'it ought to have the same powers.' In rendering needles or bars magnetic, it is found to agree with voltaic electricity, and the *direction* of the magnetism is in both cases the same, but in deflecting the magnetic needle common electricity has been found deficient. To achieve the required effect Faraday decided that some way had to be found of slowing down the electrical discharge. 'It was to the retarding power of bad conductors, with the intention of diminishing its *intensity* without altering the quantity, that I first looked with the hope of being able to make common electricity assume more of the character and power of voltaic electricity, than it is usually supposed to have.' By using a wet string as the connection between the source of common electricity (a set of jars connected to a machine like that in Figure 113), he was able to achieve his aim. 'Finally when the battery had been positively charged by about forty turns of the machine, it was discharged by the rod and the thread through the galvanometer. The needle immediately moved.' By slowing down the rate with which it was discharged Faraday showed that common electricity behaved very much like the voltaic electricity produced chemically.

Summarizing the results of other small studies Faraday covered the cases of magneto-electricity, that produced by electromagnetic induction, and animal electricity. The case of thermo-electricity, that produced by heating the junction between two dissimilar metals, was more troublesome because of the quantitatively small scale of the effect. The phenomenon had been discovered by T. J. Seebeck in 1822. Electrostatic effects, heating effects and the power to decompose solutions had not been demonstrated for this form of electricity. By a nice piece of analogical reasoning Faraday disposed of the problem this posed for his doctrine of the unity of electricities. He had already shown that the differences between common and voltaic electricity could be explained by the very high

Fig. 113: An electric generating machine. Royal Institution, London.

intensity of the former. Perhaps thermo-electricity seems different only because of its very low intensity. 'Only those effects are weak or deficient,' he says, 'which depend upon a certain high degree of intensity; and if common electricity be reduced in that quality to a similar degree with the thermo-electricity, it can produce no effects beyond the latter.'

The results of the whole study of the qualitative identity of electricities are summed up in the accompanying table, taken from Faraday's *Experimental Researches*. The sign '×' means that the effect has been experimentally established, while '+' indicates that though it has not been observed it is very probable that it does exist.

	Physiological effect	Magnetic deflection	Magnets made	Spark	Heating power	True chemical action	Attraction and repulsion	Discharge by hot air
1. Voltaic electricity	×	×	×	×	×	×	×	×
2. Common electricity	×	×	×	×	×	×	×	×
3. Magneto-electricity	×	×	×	×	×	×	×	
4. Thermo-electricity	×	×	+	+	+	+	+	
5. Animal electricity	×	×	×	+	+	×		

The identity of electricities: quantitative proof

But all this is only the first stage of the experimental series. It still remains to determine whether the electricities can be demonstrated to be quantitatively identical as well, that is whether the same amount, according to some common measure, is required to produce quantitatively identical effects. By way of preparation Faraday had to devise a common measure of quantity. To test the ability of a galvanometer to register the quantity of electricity regardless of its source and circumstances of discharge Faraday set up different sized groups of jars, but in each group stored the same amount of electricity. He determined that the amount stored was the same by using the same number of rotations of the electrical machine for each set. Though the electric tension was different in each case, greater with a smaller number of jars, less with a greater number, 'each deflected the galvanometer by the same amount on discharge'. The next step was to 'obtain a *voltaic* arrangement producing an effect equal to that just described'. By carefully adjusting the experimental arrangements Faraday designed a voltaic cell with a movable core that could be plunged into a standard acid solution, and the time it took to produce the same effect on a galvanometer as a given quantity of common electricity determined. Plunging the movable electrode in for the same time always produced the same effect. 'Hence, as an approximation, and judging from *magnetic force* only at present, it would appear that' a voltaic cell constructed in a certain way, acting for a certain time produces the same amount of electricity as an electric machine turned for a certain number of revolutions. By substituting an iodized test paper for the galvanometer the colour, depth and size of spot produced in different ways could be compared as a measure of the chemical effect. The same quantity of voltaic and of common electricity produced the same chemical effect. Faraday was able to show that the same identity held for greater quantities of electricity, measured by the time of action for the voltaic form, and the number of turns on the machine for the common. So he concludes, '. . . it is probable that for all cases, that the *chemical power, like the magnetic force is in a direct proportion to the absolute quantity of electricity* which passes.'

Subsequent developments

Though it was generally agreed that Faraday had amply demonstrated the unity of electricities, the full theoretical understanding of the experiments was lacking. It was not until the electron theory of electricity was proposed in 1897 by J. J. Thomson that the underlying explanation of these results was finally established. An electron was supposed to be a basic atom of electricity, each electron having equal electric charge. All the different methods for producing electricity were really methods for releasing streams of electrons. The number of elementary charges released determined the quantity of electricity and their rate of passage, the current. The chemical decomposition produced by the passage of electricity is an aggregate of atomic events, each of which involves the exchange of one or a small, fixed number of electrons. If this is so, the total chemical effect of the passage of electricity must be proportional to the quantity of electricity that passes, however it is produced, since it is nothing but a stream of identical electrons. Similar explanations have been found for all the common effects and common measures that Faraday collated from the work of others or demonstrated for himself.

The reasoning behind the form taken by the experimental series is quite complex. The basic principle at stake is the assumption that different modes of production yield different 'stuffs'. The principle assumed in Faraday's experimental proofs must be something like this: if several apparently different causes have exactly similar effects, both qualitative and quantitative, they must really be one and the same. To bridge the gap between the two principles, so that proof of the latter can be treated as disproof of the former, a third principle

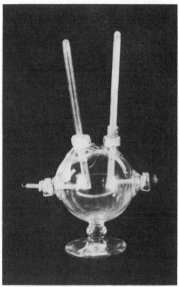

Fig. 114: A voltameter used by Faraday in his experiments. The two electrodes are at the sides of the water-filled vessel, and the electrolysed gases are collected in the two upright tubes. The volume of gases collected over a fixed period indicates the strength of the electric current. Royal Institution, London.

needs to be introduced, associating causes with active productive forces or powers. With this qualification the second principle can be treated as implying that the range of test effects is all produced by the same active power. Only so modified does the principle cast doubt on the first assumption.

Fig. 115: Faraday lecturing at the Royal Institution. *Illustrated London News* (16 Feb. 1856), p. 177.

Further reading

Faraday, M., *Experimental Researches in Electricity, 1839–1855*, 3 vols., Everyman edn., London, 1951.

Pearce Williams, L., *Michael Faraday*, London, 1965.

III

Technique

A

Accuracy and Care
in Manipulation

In the examples of experiments I have discussed so far, I have occasionally referred to the improvement of technique and to the achievement of accuracy. **J. J. Berzelius** transformed chemical experimentation, not so much by the introduction of novel apparatus or instruments, but by practising and teaching a degree of meticulousness in experimental manipulation that set quite new standards for chemical procedures.

19. J. J. Berzelius

The Perfection of Chemical Measurement

Jöns Jacob Berzelius was born at Vaversande, Ostergotland, in Sweden, in 1779. His father was a teacher, but died while Berzelius was still an infant. His mother married again, but she too died very shortly, and he was brought up by his mother's sister, 'Auntie Flora'. When she married a widower with a young family the boy was not welcome and was sent to an uncle. At twelve he was sent to school at Linkoping, where he largely supported himself by private tutoring. At this time he had a great interest in natural history. But there were troubles at the school. He was not as diligent as he should have been, and left, perhaps at the suggestion of the school authorities. In 1796 he began medical studies at Uppsala. He was very fortunate to be able, at least for a time, to learn chemistry from A. G. Ekeburg, an excellent teacher and a chemist of repute, who had discovered titanium. However, he was forced to withdraw from these studies since he could not afford the course.

Fig. 116: A daguerrotype photograph of J. J. Berzelius, taken in 1844 when he was 65. National Museum of Science and Technology, Stockholm.

The financial crisis was solved by his uncle who apprenticed him to a pharmacist, and then to a physician at a health spa. During this time he learned the techniques of quantitative analysis. Part of the mystique of the spa cures was to advertise the composition of the minerals in the spa water. At this time his interests were exclusively medical, and his doctoral dissertation, which he worked up at this time, was a study of galvano-therapy, on the uses of electricity in medicine. In 1800 he became assistant to the Professor of Surgery at Stockholm, but at about the same time began a series of chemical studies in collaboration with a young mine owner, Wittisinger. In 1805 he was appointed 'physician to the poor' in East Stockholm. He evidently continued his chemical studies during this time, since in 1807 he became Professor of Chemistry at the Karolinska Medical Institute His first work in this post concerned the composition of animal products, but he soon turned to inorganic analysis. He brought in quite new standards of rigour that transformed the chemistry of the day. In 1832 he resigned his professorship over the refusal of the National Education Commission of Sweden to grant the Institute full university status. Late in life, in 1835, he married Elisabeth Poppins. By this time Berzelius had reached international eminence, and he was created a Baron on his marriage. Despite his great fame and the honours heaped upon him he became very depressed in his old age. 'God knows', he said, 'what happens to your time once you have begun to get old. You are busy all the time, you do important things, you work, and yet when you sum it all up the result is nothing.' He died in 1848.

Analytical chemistry before Berzelius

In 1810 the study of chemistry had run up against a serious inadequacy in its empirical methods. Dalton had proposed, generalizing both brilliantly and wildly from very rough data, that when elements combined to form compounds they did so atom to atom, so to speak. Allowing for the differences in weight between the atoms of distinct elements, this combining principle leads to the hypothesis that there should be simple and fixed ratios between the amounts of constituent elements that go to form a particular compound. The basic structure of the reasoning behind all the analytical work of the period can be illustrated as follows: if sodium hydroxide is formed by the combination of clusters of atoms in which one atom of sodium combines with one of oxygen and one of hydrogen, and sodium atoms are 23 times as heavy as hydrogen atoms, while oxygen atoms are 16 times as heavy as hydrogen atoms, then in any sample of the compound the weights of sodium, oxygen and hydrogen ought to be in the ratios 23:16:1. Working backwards one ought to be able to compare a great many compounds to guess the unit weight of the atoms of elements. Then, by dividing the weights of each element found in an analysis of a compound by the relative unit weight of atoms, one can find the atomic constitution of the most elementary units of a compound. We have come to call these compound constituents 'molecules'. For instance, if one supposed that the weight of an atom of sulphur was thirty-four times that of an atom of hydrogen, and found that in a sample of hydrogen sulphide 0.04 grams of hydrogen had combined with 0.68 grams of sulphur, one could conclude by simple arithmetic that the proportion of hydrogen and sulphur atoms in hydrogen sulphide was 2:1.

Berzelius was greatly disenchanted with the inaccuracy and inadequacy of the methods of analysis in use in his day. He had started to write a textbook of chemistry for the cadets at the

Military Academy and for medical students. When he tried to bring some order and system into the existing quantitative data he found not only confusion but downright contradiction. When results were coordinated across a variety of compounds, inconsistencies appeared. The atomic theory, as elaborated by Dalton, placed strict requirements on the relationships between elements. If a given weight of an element A combines with a certain weight of element B, and the same weight of A combines with so much by weight of element C, then there should be a definite relationship between the weights of B and C when they combine. They should either be in the same ratio as they each bear to A, or some integral multiples of those weights. This allows for the possibility of there being different numbers of atoms of B and C in combination when they

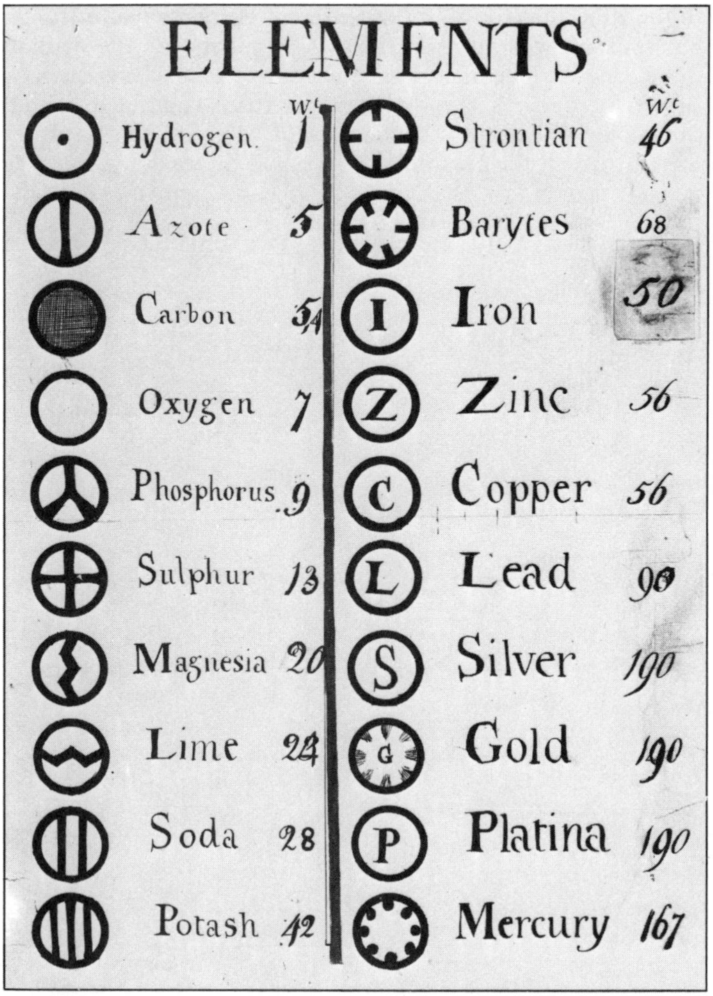

Fig. 117: Dalton's table of elements, 1806–7. Science Museum, London.

combine with each other, from when each combines with A. But Berzelius found it impossible to make existing results of measurements of relative weight fit in with these requirements. So began his obsession with precise measurement. He realized by about 1810 that progress in chemistry needed a new kind of experiment, one of meticulous, painstaking accuracy. Only then could reliable hypotheses about the atomic constitution of compounds be arrived at. He set out on ten years of devoted work to just this end.

Both Dalton (the originator of the chemical version of the atomic theory) and Wollaston (an English chemist who had pioneered quantitative chemistry) were convinced that the proportions by weight of combining substances must be integral ratios, such as 1:1 or 1:2 or 3:2 and so on. This followed directly from the atomic theory, along with the assumption that the atoms of different elements had different but constant weights. Berzelius was familiar with the work of these English chemists, and he knew also of Gay-Lussac's successful demonstration that when gases combined chemically, they did so in integral ratios of volumes, so that water was formed by the combination of two volumes of hydrogen to one of oxygen. At that time, it must be remembered, the

Fig. 118: Some examples of Berzelius's apparatus. Illustration from the original Swedish edition of the *Treatise on Chemistry*, vol. III, plate I.

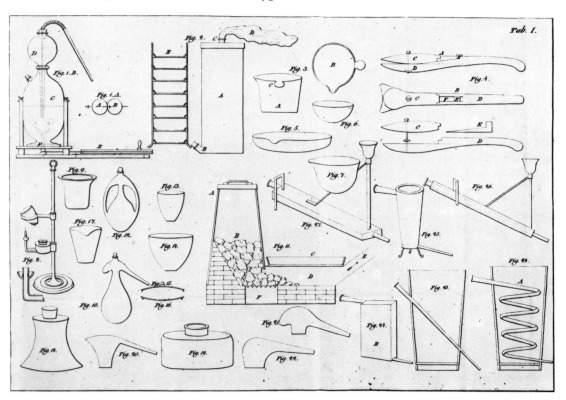

familiar distinction between atoms and molecules had not been formulated. Contemplation of all these matters led Berzelius to the conviction that equal volumes of permanent gases (those which could not then be liquified) must, at the same temperature and pressure, contain equal numbers of atoms. There must then be a relation between the integral ratios of volumes and the integral ratios of weights, revealed in studies of chemical combination. This notion was later to be incorporated in more refined form into chemistry as Avogadro's Hypothesis. Incomplete though these ideas proved to be they were sufficient to give Berzelius a powerful enough theoretical basis for his purposes, a theory which foretold that combining weights must be in integral proportions. This enabled him to formulate the idea of a 'correct' measurement.

A measurement was correct when it gave integral proportions, for that was required by the atomic theory. In his autobiography Berzelius notes, 'many times I had to repeat my analysis with different methods to find that method which was most certain to give the correct result', that is the result in accordance with the atomic theory. Berzelius did not *discover* that the elements combined in integral proportions. By assuming that that was indeed the way they must combine he corrected and improved and adjusted his experimental technique until his results were in accordance with this principle.

The analytical programme

The secret of his success was a kind of perfectionism, an obsession with accuracy. 'My first attempts in this were not successful,' he says. 'I still had no experience regarding the great accuracy that was needed, nor how a greater accuracy could be obtained in the final results.' The answer to these troubles lay in attention to detail. Equipment had to be designed so that there was as little loss of material as possible. In reactions which required pouring the vessels had to have lips that discharged the very last drop. Filter papers not only had to have a standard residue of ash, but it was advisable to wet them before they were to be used, to prevent some of the substances dissolved in the solute being absorbed by the fibres of the paper. But above all the manipulative technique had to be precise. It consisted in 'observing a large number of small details which, if overlooked, often spoil several weeks of careful work'.

Atomic weight determinations depended on two things. It was necessary to know the relative numbers of atoms of different elements in compounds, for instance, whether an oxide was ZnO or ZnO_2 or Zn_2O and so on. It was also necessary to know the equivalent weights of the elements so combined. Knowing the relative weights of zinc and oxygen in zinc oxide, and knowing that the atomic composition of the oxide is one atom of zinc to one of oxygen, the relative weights are the relative weights of the atoms of each element. Standardization was achieved by referring all weights to that of oxygen.

The basic method perfected by Berzelius involved oxygen compounds. These were more common and much more easily handled than the hydrogen compounds that had been favoured in England. But the use of hydrides was fairly common, and Berzelius gives his results in terms of both the hydrogen and the oxygen standard. One could either start with a given amount of a metal and form the oxide, or start with the oxide and reduce it to the metal with hydrogen. The particular technique chosen by Berzelius depended on the ease of manipulation and the possibility of error.

The reasoning is very simple. The combining ratio is

$$\frac{\text{Weight of oxide} - \text{weight of metal}}{\text{weight of metal}}$$

If the atomic proportions are known by comparison with other analyses it is a simple matter to calculate the ratios of the atomic weights. For instance, if the oxide consists of two atoms of oxygen to one of the metal then the above ratio must be divided by 2.

Here is Berzelius's description of the steps involved in finding the atomic weight of chlorine relative to oxygen and to hydrogen. In his *Treatise on Chemistry*, volume V, he says, 'I established its [chlorine's] atomic weight by the following experiments: (1) From the dry distillation of 100 parts of anhydrous potassium chlorate, 38.15 parts of oxygen are given off and 60.85 parts of potassium chloride remain behind. (Good agreement between the results of four measurements.) (2) From 100 parts of potassium chloride 192.4 parts of silver chloride can be obtained. (3) From 100 parts of silver 132.175 parts of silver chloride can be obtained. If we assume that chloric acid is composed of 2 Cl and 5 O, then according to these data 1 atom of chlorine is 221.36. If we calculate from the density obtained by Lussac, the chlorine atom is 220 [relative to the atomic weight of oxygen]. If it is calculated on the basis of hydrogen then it is 17.735.'

The simplicity of the reasoning and the need for careful manipulation are vividly illustrated in this passage. To get to the final ratio between the element in question (chlorine) and the standard (oxygen), several different ratio determinations have to be gone through, each of which must be as accurate as possible. Berzelius's result is in good agreement with modern determinations, but for one thing. It is only half the modern value. The reason lies in the way the hydrogen standard was computed. Without the distinction between atoms and molecules it was natural to think of hydrogen as a monoatomic gas. If one thinks of the ultimate particles of hydrogen as atoms, single Hs, when as we now think they are really molecules, H_2s, pairs of atoms, one will be inclined to take $2H = 1$ as the standard, and this is just what Berzelius did. Correcting the value gives us an atomic weight for chlorine of 35.47, relative to hydrogen.

From the point of view of scientific method it is worth noticing Berzelius's devotion to the 'intensive design'. There are two ways of gaining general knowledge by experiments. One can study a great many samples and then find their typical properties by some sort of averaging. This is called the 'extensive design'. Or one can take one, or at most a very few cases, and assume that they are typical. Their properties will then be the defining properties of all samples similar to them. This is called the 'intensive design'. As MacNevin says of Berzelius, 'The selection of the proper method of analysis seemed far more important to him than the frequent repetition of the measurement common today . . . he seldom repeated any of it once completed and was ready to defend its reliability.'

By 1818 Berzelius was ready to announce the atomic weights of 45 of the 49 known elements. Throughout his life he continued to improve and extend these results.

Fig. 119: Liebig's laboratory in 1842. This engraving shows a typical nineteenth-century chemistry laboratory. Deutsches Museum, Munich.

Berzelius was not just a superb experimenter. He developed, in much the same way as had Davy, an electrical theory of chemical combination, but with a more detailed and precise form. Soderbaum, quoted in Jorpes, gives Berzelius as saying, 'Atoms contain both types of electricity, these being placed at different poles in them, but one type is dominant. Affinity is due to the effect of the electrical polarities of the particles. Thus, all compounds are composed of two parts, these parts differing in the nature of their electricity, and are bound together by attraction. All compounds can therefore be divided into two oppositely charged parts irrespective of the number of elements from which the compound is composed.'

This was a powerful theory. It did very well for inorganic compounds, but the discovery that chlorine could be substituted for hydrogen, atom by atom, in organic compounds, brought it into temporary disfavour (and brought the discoverer of substitution, Liebig, into permanent disfavour with Berzelius!). If hydrogen is electropositive, any atom which takes its place in another compound should, according to Berzelius's theory, also be electropositive, since it would be held in place by the negative charge on the other component. But chlorine is electronegative. The fact that chlorine can be substituted for hydrogen in many hydrocarbons, for instance, seems directly to contradict the Berzelian theory.

Finally, one must mention the vast influence on both chemistry and chemists exerted by Berzelius as the author of the chemical Yearbook (*Jahresbericht*) in which for 27 years he summarized, commented upon and criticized advances in chemical knowledge and technique from all over Europe.

Atomic Weights after Berzelius

Berzelius's methods depended on the accuracy with which he was able to infer the proportions of each kind of atom in a compound. He was also able to utilize a more direct method for spot-checking some atomic weights. The technique had been perfected by Dulong and Petit. They had been exploring the consequences of Dalton's idea that the heat capacity of the atoms of all gases was related to their relative size. They found that his hypothetical figures were very much in error. In the course of this work they noticed an important relation between atomic weight and specific heat; that is the amount of heat required to raise the temperature of a standard mass of a substance by a standard amount. This relation, verified only for solid substances, later came to be known as their law of atomic heat. It turned out that the product of atomic weight and specific heat of an element was a constant. With the help of Regnault they checked Berzelius's results, and found that some of his figures should be doubled and others halved, for instance the atomic weights of silver and sulphur were wrong. Unfortunately, though their law did allow some direct check on Berzelius's results, it had some exceptions, and was not a wholly reliable guide. But gradually the combination of more and more exact chemical knowledge, a clearer idea of the difference between atoms and molecules, and further refinement of direct measurement techniques cleared up most of the anomalies during the nineteenth century. But it remained to explain why the measured atomic weights were not whole numbers.

In 1886 Crookes first put forward the idea that the elements as we know them may be mixtures of yet more elementary substances, the masses of the atoms of which were related to the mass of oxygen atoms in integral proportions, more or less as Prout had suggested. But this idea was not experimentally verified until F. W. Aston developed the mass spectrograph. By developing J. J. Thomson's magnetic and electrical field equipment by which he had studied the physical properties of

Fig. 120: F. W. Aston with his mass spectograph in the Cavendish Laboratory, Cambridge.

electrons (see Experiment 16), Aston was able to separate atoms of the same electrical charge but different mass. Previously these had been taken to be all of the one kind, *the* atoms of neon, say. The confusion had arisen because it turned out that the chemical behaviour of atoms was largely determined by their electrical properties and very little by their mass. Aston showed that elements of even atomic number, that is having an even number of electrons in their atoms, tended to form two isotopes (as they came to be called). These each had nearly integral weights, and the traditional atomic weight, so carefully computed by Berzelius, was the result of a mixture of isotopes. Different elements were found in nature to be made up of different proportions of their isotopes. This was why even that paragon of accuracy, Berzelius, had found the atomic weight of chlorine to be that awkward number 35.47 (adjusted to the modern hydrogen standard).

In these experiments we see the refinement of a measuring technique. But 'refinement' is correlative to the idea of 'correct result'. Without some prior conception of how things ought to go, we can have no idea of a correct or an incorrect result. With the help of atomic theory Berzelius was able to anticipate his experimental results, using theory to correct experiments.

Further reading

Berzelius, J. J., *Essai sur la théorie des proportions chimiques et tables synoptiques de poids atomiques*, Paris, 1819.

Berzelius, J. J., *Traité de chimie* [*Treatise on Chemistry*], 8 vols., Paris 1829–33 (this edition, translated from the Swedish, is the most accessible to English readers).

Jorpes, J. E., *Jac. Berzelius: His Life and Work*, Royal Swedish Academy of Science, Stockholm, 1966.

MacNevin, W. M., 'Berzelius, pioneer atomic weight chemist', *Journal of Chemical Education*, 1954, pp. 207–10.

Szabadvary, F., *History of Analytical Chemistry*, transl. G. Svehla, Oxford, 1966, ch. VI, sections 2 and 3.

B

The Power
and Versatility of Apparatus

Manipulative care is but one side of the story of technique. The other is to be told in terms of the ingenuity, fruitfulness and power of the apparatus with which experiments are finally actually conducted. It is only too easy to think of an experiment in purely logical terms, particularly if one's knowledge of experimentation comes from reading the finished products of an investigation, the scientific paper or textbook. Sometimes a whole new field is opened up by the invention of an apparatus of great power and versatility. One of the most successful of such pieces of apparatus was the equipment developed by **Otto Stern** and H. Gerlach for the production and study of molecular beams.

20. Otto Stern

The Wave Aspect of Matter and the Third Quantum Number

Fig. 121: Otto Stern.

Otto Stern was born in Schrau, in the province of Upper Silesia, formerly part of Germany, in 1888. His father was a very prosperous grain merchant and miller. The economic security of the family greatly influenced Stern's scientific career. He was the eldest of five children. His primary and secondary education was at Breslau (now in Poland). From 1906 he wandered about the German universities in the fashion of the day, working at Freiburg, Munich and Breslau. As a young man of independent means he was even more free than the majority of German students to indulge his interests, and to work on projects that were not directly related to a career. It was his interest in thermodynamics that drew him back to Breslau where there was a school of physical chemistry, centred on thermodynamic properties of chemical relations. He took a PhD in physical chemistry there in 1912.

In that year he came under the influence of Einstein, doing post-doctoral work with him in Prague and then moving with Einstein to Zurich in 1913. It was Einstein's molecular studies, rather than his relativity theory, that interested Stern. In 1914 he met Max Born, and began to work with him, having been licensed as a *Privatdozent*, an unsalaried lecturer at the university, in that year.

Stern served in the German Army throughout the First World War, but contrived to continue scientific work. For a while he was a meteorologist in Poland, but in the last year of the war was among the scientists seconded to work in Nernst's laboratory in Berlin.

It was after the war that he developed his molecular beam methods for studying free atoms, on the analogy of light beams. His beams of atoms, upon which the experiment described in this section depended, were the basis of his demonstration of the wave-like properties of matter, which, in classical physics, had been assumed to be wholly particle-like. In 1923 he moved to Hamburg to his own laboratory. With new and greater facilities he was able to develop the molecular beam methods still further, and it was there that the actual demonstrations of the wave aspects of matter were achieved.

Stern left Germany in 1933 under the threat of the Nazi regime, and settled in the United States, working at the Carnegie Institute. Unlike some of those driven out by the Hitler government he never did strike the same vein of productive work as he had had to abandon in Germany. He was awarded the Nobel Prize in 1943. In 1946 he retired to Berkeley, California, and died there in 1969.

The context of the experiments

The discoveries which flowed from Stern's development of molecular beam apparatus were relevant to problems which had been formulated almost at the very time that Stern was at work on testing hypotheses of his own contriving. These hypotheses turned out to bear directly on main-stream physics. The electron theory of the atom had been developed by Neils Bohr from hints that had emerged from Rutherford's demonstration of the nucleated form of atoms. If both the positive charge and the bulk of the mass of an atom were concentrated in a small central region or nucleus, then it would seem reasonable to suppose that the remaining mass and the balancing negative charge would be concentrated in the periphery. A natural step was to treat electrons as small charged bodies and to imagine them revolving around the nucleus in some kind of planetary motion. This idea suggested a number of consequential concepts, and problems. If electrons had *orbits*, how were these orbits arranged in space? If electrons *moved* in those orbits, what was their angular momentum, the impetus of their rotary motion? If the orbits lay in a plane or planes, as do the planetary orbits of the solar system, how are those *planes* related to each other? And finally, if the electrons were small charged bodies they could rotate or *spin* about their own axes. What were the directions of their spin?

Bohr proposed that the emission of light by hot gases and solids should be explained by supposing that electrons changed their orbits, and lost energy in the process, energy which was emitted as light. But it had been known for some time that the light emitted by incandescent substances appeared not as a continuous spectrum but in distinct wavelengths. To explain this Bohr supposed that the electrons could take up only some of the mechanically possible orbits. These were represented by what came to be called the 'principal quantum number', $n = 1$,

Fig. 122: Walther Gerlach.

2, 3 . . . The term 'quantum' referred to the discrete 'packet' of energy released by the jump from one fixed orbit to another, given out as light of a definite wave length.

Further study of atomic phenomena suggested that there should be a second quantum number. Electrons seemed not to have just any angular momentum, but to orbit only with certain definite velocities. This feature of the structure of atoms was represented by a letter l. l could be related to n, since the admissible angular momenta were represented by only those integral values of l that lay between 0 and $n-l$.

Would the same kind of feature be revealed for the other major properties of planetary motions, the orientation of the orbital planes and the direction of spin? Would it turn out that the orbits of electrons could be only in some definite planes? Would spin too turn out to be, as physicists came to say, 'quantized'? To represent these possibilities, two further quantum numbers were proposed; m was introduced to represent 'space quantization', the permissible angles that the planes in which electrons orbited could make with some fixed plane, such as a magnetic field imposed from outside. A fourth quantum number, s, was added to represent the possibility that electrons could spin only clockwise or anticlockwise relative to one fixed axis. This has come to be called 'spin up' or 'spin down'. It seemed that all the properties of the electrons that mattered could be expressed by the use of these four quantum numbers.

One might ask why the architecture of atoms should be represented by numbers. It is very easy to see why this should be in the case of the layout of the orbits of electrons around the heavy nucleus. If, in the lowest orbit, an electron has energy e, an electron in the next possible orbit will have an energy $2 \times e$, in the next $3 \times e$, and so on. The numbers 1, 2, 3 and so on are the values of n, the principal quantum number, and define the possible orbits for electrons. The first use of the Stern–Gerlach apparatus that I shall describe was directed to testing whether the idea of introducing a third quantum number, space quantization, was right.

Fig. 123: A representation of the meaning of the four quantum numbers.

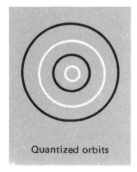

Quantized orbits

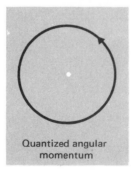

Quantized angular momentum

Quantized orbital planes

Quantized spins

These quantum numbers came from considering electrons as particles and seeing what changes to the traditional principles of mechanics needed to be made to accommodate their peculiar behaviour. But it was already well established that beams of electrons behaved in a very odd way when they interacted with each other. They showed interference effects as if they were waves. They could be diffracted too. This behaviour suggested another radical change in the standard physics of the time. Perhaps there were rules which could represent the wave-like characteristics and the particle-like characteristics in some unified way. De Broglie proposed that two aspects of electron behaviour which would, in the traditional physics, have been treated as distinct and unrelated phenomena, should be combined by allowing the following transformation: The wave length, λ, and the mechanical aspects of a particle of mass, m, and velocity, v, would be related by the formula

$\lambda = \mathrm{h}/mv,$

where h is Planck's constant. What if this relation were true in general, not only for electrons, but for atoms, cannon balls, planets and so on? Clearly the molecular beams of Stern would provide an ideal testing ground for this idea. If a beam of molecules could be diffracted, say, then the de Broglie principle could be taken much more seriously as a general physical law. Molecular beams could be used to test this idea.

The experiments with molecular beams

The importance of Stern's experiment is obvious in the context of the development of physics. But it illustrates another point of interest in studying experiments, namely the power of some techniques to provide answers to a number of different questions, sometimes not fully formulated when the equipment was first developed.

The Stern–Gerlach apparatus is based upon the permutation of three pieces of sub-equipment; there is a device for preparing a beam of molecules (or atoms) all with nearly the same velocity; then there is an arrangement for producing an intense and steeply 'graded' magnetic field, changing greatly in intensity over a very small distance; and finally there is the use of metallic crystals to provide gratings suitable for the diffraction of wave motions of wave-lengths which bodies of atomic dimensions would have, if the De Broglie laws defining the characteristics of associated waves were correct.

To prepare a beam of suitable atoms Stern (with his assistant, Gerlach) used a small crucible, at high temperature, into which samples of the appropriate substances were passed. A narrow slit opened into a vacuum and so, with the heat to

Otto Stern

provide the thrust, a beam of atoms was produced. But these were of all sorts of energies (velocities). To stop all but those of a very narrow range of velocities they adapted the idea used by Fizeau to measure the velocity of light. The problem is created by the very high speeds of the atoms. If two wheels are fitted on the same axis, both with slits in them, and set to contra-rotate, only those atoms which take just the time for one slit to be replaced by another, to cross between the wheels, will be able to pass right through. So all atoms which do succeed in running the gauntlet of the two wheels will be of roughly the same speed.

Similar ingenuity was shown in their development of the second part of the equipment, the high-density magnetic field. The atoms are passing very swiftly through the system, so for

Fig. 124: The arrangement to separate out atoms moving at the same speed.

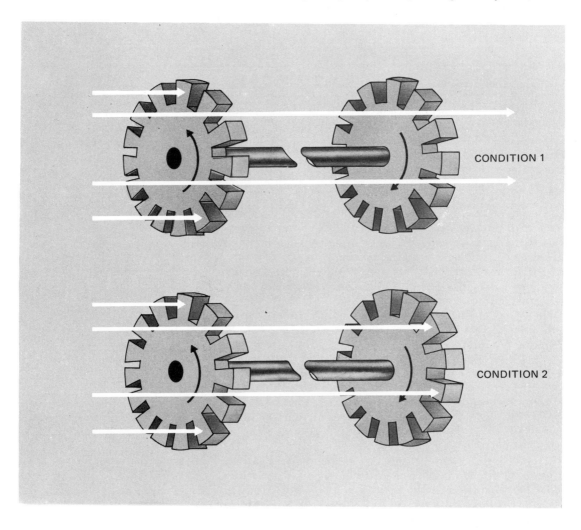

CONDITION 1

CONDITION 2

there to be any discernible effect, say a splitting of the beam by virtue of the mysterious property of space quantization, the field must be very concentrated. By forming one pole as a knife edge, and the other as a groove and contriving that the beam sped along the narrow gap between, the maximum effect should be produced.

The equipment required to detect the diffraction of matter did not require any particularly novel arrangements. The basic technique for using crystals to diffract electrons had been developed by Elsasser and refined by Davisson and Germer in 1927. So when Stern adapted his molecular beams to study matter diffraction in collaboration with Estermann in 1929 or thereabouts, they simply had to borrow the technique. In the event they used lithium crystals for the diffraction target, and a

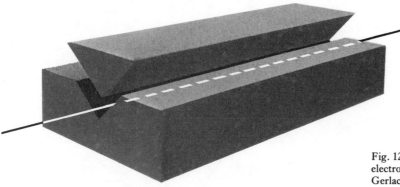

Fig. 125: The pole pieces of the electromagnet inside the Stern–Gerlach apparatus.

small chamber into which diffracted helium atoms could be collected and their quantity measured by detecting minute pressure changes.

Once all this had been thought out and the equipment set up the experiments themselves were extraordinarily direct, a sure sign of genius in an experimenter. Moving electric charges create a magnetic field. Orbiting electrons are electric charges, and, according to the atomic theory, they are moving. So they should create magnetic fields. If all the electrons in each atom trace out their orbits in only one of the possible planes allowed by the quantum theory, then the associated magnetic field of each atom will be associated with that plane. So when an external magnetic field is allowed to affect the atoms they should take up quite definite orientations to that field, depending on their internal magnetic fields.

If there were space quantization, as I have described it in the previous subsection, the orientations of the little magnets, as we are now to imagine the atoms, will not be at random, 'all round the clock', so to speak, but in the particular case Stern was investigating, at *two* distinct angles to the external magnetic field, as predicted from the calculation for the third quantum number. Each orientation corresponds to one of the planes in which electrons might orbit, and so if we look at the image of the beam with the field switched off we should find a single blur on the photographic plate. But when the field is switched on the beam should separate into two, those atoms quantized one way with respect to the plane of their orbiting electrons moving away one way, and those quantized in the other angle moving off in the other way. And that is exactly what Stern found.

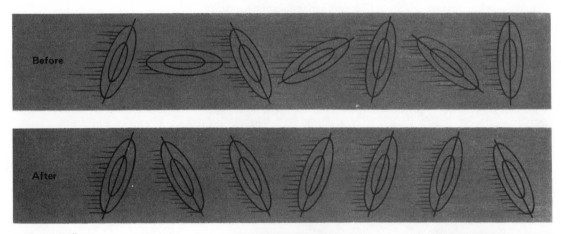

Above Fig. 126: The effect of switching on the magnetic field on the orientation of atoms, showing the atoms being 'quantized' into one or other of only two planes.

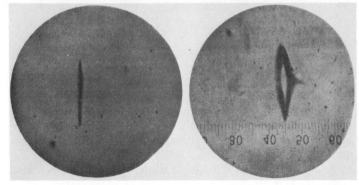

Right Fig. 127: The separation of the beams. From an article by W. Gerlach in the *Zeitschrift für Physik*, vol. 9 (1924), p. 350.

The demonstration of the wave nature of matter was equally direct. Experiments by Elsasser, and Davisson and Germer had shown that electrons could be diffracted, apparently a proof of their having some wave-like modes of behaviour. But this was rather a special case and could not be taken as evidence for a general matter–wave equivalence. Electrons no doubt had the peculiar character, behaving like material particles in one kind of set-up and like waves in another. But helium atoms are relatively lumpish and commonplace bits of matter. If they showed diffraction effects, then de Broglie's idea for a

Fig. 128: General view of the apparatus. From an article by W. Gerlach in the *Journal de la Physique et le Radium*, vol. 10 (1929), plate I.

thoroughgoing equivalence between matter and waves was much more firmly established.

For this range of experiments the equipment was permuted. The beam-producing equipment with its contra-rotating wheels was used to be sure that all the atoms were at the same speed. The beam-producer was coupled with a lithium crystal for a target, and with a detector to measure the angle through which atoms were diffracted. If they were being mechanically reflected, behaving as a stream of particles, as tennis balls do when reflected off a volley board, then the angle of reflection would be about the same as the angle of incidence. But if they were being diffracted, behaving like a wave, then there should be a spread of diffracted atoms, like a diffracted wave front. Again, Stern had devised an experiment that gave a simple and direct answer to the question.

Not surprisingly, though the result would have been almost unintelligible to physicists of a generation or two before, Stern and Estermann found just the distribution, or spread, that they were expecting. The pressure in their little collecting 'jar' rose to a peak in just the way it should if the beam of helium atoms was behaving like a wave.

Developments after Stern and his collaborators

Molecular beam laboratories proliferated as Stern's pupils began to have pupils of their own. The equipment has been greatly refined, new ways of detecting the effects of the beams on different kinds of materials have been developed. But it can be said that in this field, though much ingenious work has been done, the essentials were achieved by its originator, Otto Stern himself.

Even so convincing a series of demonstrations of the particle nature of subatomic matter as were given by Thomson and Rutherford may be upset by new concepts. De Broglie's extraordinary generalization of his rules to all material things suggested the possibility of wave-like effects even for relatively massive bodies like whole atoms. The consequences of the discovery of just the kind of effects his generalization suggested have not yet been fully absorbed into the metaphysics of natural science.

Further reading

The original papers are in German and are to be found in *Zeitschrift für Physik*, 9 (1922), pp. 349–52, a paper by W. Gundlach and O. Stern; and in *Zeitschrift für Physik*, 61 (1930), p. 95, in a paper by I. Estermann and O. Stern.

Estermann, I. (ed.), *Recent Research in Molecular Beams*, a collection of papers dedicated to Otto Stern, on the occasion of his 70th birthday, New York–London, 1959, pp. 1–7 ('Molecular Beam Research in Hamburg, 1922–1933').

Schonland, B., *The Atomists, 1805–1933*, Oxford, 1968, ch. 11.

General bibliography

Bloor, D., *Knowledge and Social Imagery*, London, 1976. This is one of the most convincing attempts yet to link the way science develops to the social forces of a historical period.

Crombie, A. C., *Medieval and Early Modern Science*, volumes I and II, New York, 1959. Prevailing mythology suggests that science began with Copernicus. In this detailed but readable survey the medieval achievements are well brought out.

Gillispie, C. C., (ed.), *Dictionary of Scientific·Biography*, New York, 1970–6; supplement 1978. This contains detailed studies (and bibliographies) of the life and works of all important scientists.

Harré, R., *Philosophies of Science*, London, 1976. Throughout history two major views of science have competed, positivism – the theory that science is concerned only with what can be observed – and realism – the idea that scientists can enter a realm beyond experience. This work compares and contrasts these grand positions.

Harré, R. (ed.), *Scientific Thought, 1900–1960*, Oxford, 1969. In this collection each major scientific field is described by one of the scientists who was at the forefront of advances in the period.

Holton, G., *Thematic Origins of Scientific Thought: Kepler to Einstein*, Cambridge, Mass., 1973. A detailed and convincing study of how scientists think.

King, L. S., *The Growth of Medical Thought*, Chicago, 1963. This study shows how changing conceptions of disease affected the interpretations placed upon medical experience.

Latour, B., and Woolgar, S., *Laboratory Life*, Beverly Hills and London, 1979. The authors look at scientific laboratories as if they were exotic localities inhabited by strange tribes, to try to make sense of what scientists actually do in their day-to-day activities.

Leicester, H. M., and Klickstein, H. S., *A Source Book in Chemistry*, Cambridge, Mass., 1965. Selections from the original publications with short commentaries cover the whole history of chemistry.

Losee, J., *A Historical Introduction to the Philosophy of Science*, London, 1972. Theories of the very nature of science as a mode of knowing about the world have changed along with the changes in science itself. These changes are systematically treated in relation to the history of philosophy and of science.

Magie, W. F., *A Source Book in Physics*, Cambridge, Mass., 1963. Most of the important discoveries in physics are represented in this collection, made up in the same fashion as the chemistry source-book.

Medawar, P., *The Art of the Soluble*, London, 1967. This collection of essays contains a marvellous account of the uses of the imagination in scientific discovery.

Singer, C., *A History of Biology*, London and New York, 1960. Emphasis is placed on the interaction between changing ideas of life and living function and the slow growth of anatomical, physiological and historical knowledge of living forms.

Singer, C., *A Short History of Scientific Ideas to 1900*, Oxford, 1959. This is still the most clearly written and comprehensive history of science available.

Index

Albertus Magnus, Saint (*c*.1200–80) German scholastic philosopher, 95

Amagat, E. H. (1841–1915) French physicist, 22, 89, 91

Ampère, A. M. (1775–1836) French physicist, 52

Andrews, T. (1813–85) English physicist, 22, 89

Ardrey, R., scientific popularizer, 73

Aristotle (384–322 BC) Greek philosopher and scientist, 11, 23, 31–8, 57, 95

Aston, F. W. (1877–1945) English scientist, 209–10

Atwood, G. (1746–1807) English physicist and mathematician, 82

Avicenna (Ibn Sina) (980–1037) Persian physician and philosopher, 36

Avogadro, Amedeo (1776–1856) Italian physicist, 205

Bacon, Francis (1561–1626) English philosopher and man of letters, 10, 15, 20, 21, 22

Bayliss, W. M. (1860–1924) English physiologist, 47

Beaumont, W. (1785–1853) U.S. army surgeon and physiologist, 39–48

Becquerel, A. H. (1852–1908) French physicist, 114

Beddoes, T. (1760–1808) British chemist and physician, 163

Bell, Alexander Graham (1847–1922) U.S. inventor and physicist, 124

Berzelius, J. J. (1779–1848) Swedish chemist, 19, 155, 201–10

Bohr, N. (1885–1962) Danish physicist, 112, 213

Born, M. (1882–1970) German physicist, 212

Boscovich, R. J. (1711–87) Serbian theoretical physicist, 163

Boussingault, J. B. (1802–87) French chemist, 64

Boyle, Robert (1627–91) English scientist, 11, 22, 23, 28, 60, 83–91, 114

Bradwardine, Thomas (*c*.1290–1349) English mathematician, 77

Broglie, L. V. de (1892–), French physicist, 215, 220

Burroughs, W. (1855–98) American inventor, 49

Cavendish, H. (1731–1810) English chemist and physicist, 155, 161

Chadwick, Sir James (1891–1974) British physicist, 112

Chandler, B., 39

Clausius, R. J. E. (1822–88) German theoretical physicist, 89

Clavius, Christopher (1537–1612) German astronomer and mathematician, 17, 18

Crick, F. (1916–) British molecular biologist, 139, 142

Crookes, Sir William (1832–1919) English physicist, 209

Curie, Pierre (1859–1906) and Marie (1867–1934) French physicists, 114, 170

Dalton, John (1766–1844) English chemist and physicist, 202, 203, 204, 208

Darwin, Charles (1809–82) English naturalist, 21, 66, 68

Davaine, C. (1812–82) French bacteriologist, 105

Davisson, C. J. (1881–1958) U.S. experimental physicist, 217, 219

Davy, Sir Humphry (1778–1829) English chemist, 12, 24, 163–70, 174, 191, 208

Descartes, René (1596–1650) French philosopher, 99, 126, 183, 188

Dulong, P. L. (1785–1838) French chemist and physicist, 208

Dumbleton, John (d. *c*.1350) English philosopher, 77

Einstein, Albert (1879–1955) German physicist, 11, 133, 212

Ekeburg, A. G. (1767–1813) Swedish chemist and mineralogist, 201

Elsasser, W. M. (1904–) German born U.S. geophysicist, 217, 219

Estermann, I. (1900–) American physicist, 217

Euler, L. (1707–83) Swiss mathematician, 188

Fabricius ab Aquapendente, Hieronymus (1537–1619) Italian anatomist and embryologist, 36

Faraday, Michael (1791–1867) English chemist and physicist, 9, 12, 14, 53, 56–7, 147, 165, 172, 174, 191–8

Fitzgerald, G. F. (1851–1901) Irish physicist, 132

Fizeau, H. L. (1819–96) French physicist, 126

Frisch, Karl von (1886–) Austrian zoologist, 66, 73

Galen (*c*. AD 130–*c*. 200) Greek physician, 36

Galilei, Galileo (1564–1642) Italian mathematical physicist, 76–82, 91, 182

Galilei, Vincenzo (*c*.1520–91) Italian mathematician and music theorist, 76

Galvani, Luigi (1737–98) Italian physiologist, 165

Gay-Lussac, J. L. (1778–1850) French chemist and physicist, 204

Geiger, J. W. (1882–1945) German physicist, 112

Gerlach, Walther (1889–) German physicist, 27, 214, 215

Germer, L. H. (1896–) U.S. physicist, 217, 219

Gibson, J. J. (1904–) American psychologist, 12, 147–53

Gilbert, W. (1544–1603) English scientist, 11, 146, 188

Giles of Rome (Aegidius Romanus) (*c*.1245–1316) Italian theologian and philosopher, 36

Grew, Nehemiah (1641–1712) English plant anatomist and physiologist, 11, 23, 60, 61

Guericke, Otto von (1602–86) German physicist, 85

Hales, Stephen (1677–1761) English plant physiologist and chemist, 11, 23, 57–64, 157, 158

Hartlibb, Samuel (1599–1662) English reformer and agriculturalist, 83

Harvey, William (1578–1657) English physician, 11, 58

Hayes, W. (1913–) English molecular biologist, 142

Heinroth, O. (1871–1945) German ornithologist, 65, 67, 68

Hellman, B., 65

Helmholtz, H. L. F. (1821–94) German physiologist and physicist, 172, 174

Helmont, J. B. van (1579–1644) Flemish physician, 39–41, 103

Hermes Trismegistus, 9

Hertz, H. R. (1857–94) German physicist, 174, 175, 176

Heytesbury, William (c.1310–80) English logician and mathematician, 77

'Hippocrates' (fl. 400 BC) Greek physician, 32, 38

Hittorf, J. W. (1824–1914) German physicist, 174

Hooke, Robert (1635–1703) English experimental physicist, 22, 28, 60, 86, 186

Hunter, I. M. (1915–75) English physicist, 112

Huyghens, C. (1629–95) Dutch scientist, 188

Ingenhousz, J. (1730–99) Dutch physician, 64

Jacob, F. (1920–) French molecular biologist, 137–46

Jenner, E. (1749–1823) English physician, 104, 105

Kant, Immanuel (1724–1804) German philosopher, 163

Kepler, J. (1571–1630) German astronomer, 76, 81, 182

Koch, R. (1843–1910) German physician, 110

Lavoisier, A. L. (1743–94) French chemist, 12, 155–62, 163, 164, 169, 170

Lenard, P. E. (1862–1947) German physicist, 178, 179

Liebig, Baron Justus von (1803–73) German chemist, 208

Lister, Joseph (1827–1912) English surgeon, 105

Locke, John (1632–1704) English philosopher, 83

Lorentz, H. A. (1853–1928) Dutch physicist, 132, 133

Lorenz, K. (1903–) Austrian zoologist, 65–74

Macfarland, D. J. (1938–) English zoologist, 73

Malpighi, M. (1628–94) Italian physiologist, 61

Marci, J. M. (1595–1667) Bohemian physician, 185, 187

Marsden, P. L. (1924–) English physicist, 112

Maxwell, J. C. (1831–79) British physicist, 134

Mayow, J. (1640–79) English chemist and physiologist, 59, 64, 157, 158, 159, 161, 162

Mendel, G. J. (1822–84) Austrian botanist, 137

Michelson, A. A. (1852–1931) U.S. physicist, 11, 25, 28, 124–34

Monod, J. L. (1910–76) French zoologist, 137

Morley, E. W. (1838–1923) U.S. chemist and physicist, 11, 25, 28, 124–34

Müller, J. P. (1801–58) German physiologist, 39

Newton, Sir Isaac (1642–1727) English physical scientist and mathematician, 11, 14, 81, 126, 133, 182–9

Norman, Robert (fl. c.1590) English scientist, 11, 49–56, 91, 146, 188

Oliphant, Sir M. L. E. (1901–) Australian-born physicist, 112

Pasteur, Louis (1822–95) French chemist and microbiologist, 11, 14, 18, 102–10

Pavlov, I. P. (1849–1936) Russian physiologist, 46

Perrin, J. B. (1870–1942) English physicist, 174, 175

Peter of Alvernia (died 1304) French scholastic philosopher, 95

Petit, A. T. (1791–1820) French physicist, 208

Plucker, J. (1801–68) German mathematician and physicist, 174

Popper, K. R. (1902–) British philosopher of science, 18

Poynting, J. H. (1852–1914) British physicist, 171

Priestley, Joseph (1733–1804) English chemist, 155, 157–8, 159, 161, 162

Prout, W. (1785–1850) British chemist, 19, 209

Regnault, H. V. (1810–78) French chemist and physicist, 208

Roemer, O. (1644–1710) Danish astronomer, 126

Rossignol, A. M., 109

Rutherford, E. (1871–1937) British physicist, 12, 111–22, 172, 179, 213, 220

Schuster, A. (1851–1934) British mathematician and physicist, 171

Schwann, T. (1810–82) German physiologist, 104

Seebeck, T. J. (1770–1831) German physicist, 195

Snell, W. (1591–1626) Dutch astronomer and mathematician, 100

Soddy, F. (1877–1956) British physicist, 111

Starling, E. H. (1866–1927) English physiologist, 47

Stern, O. (1888–1969) U.S. physicist, 27, 212–20

Stukeley, W. (1687–1765) English antiquary, 58

Swineshead, Richard (fl. 1337–48) English philosopher, 77

Theodoric of Freibourg (d. c.1311) German philosopher and scientist, 11, 28, 57, 91, 93–100, 183

Theophrastus (c. 370–c. 287 BC) Greek philosopher, 59

Thomson, J. J. (1856–1940) English physicist, 12, 111, 122, 171–80, 197, 220

Tinbergen, N. (1907–) Dutch zoologist, 66, 73

Torricelli, Evangelista (1608–47) Italian physicist and mathematician, 84–5

Toussaint, 106, 108

Trutfetter, Jodocus, 100

Volta, A. (1745–1827) Italian physicist, 165

Wallis, J. (1616–1703) English mathematician, 83

Ward, S. (1617–89) English mathematician, 83

Watson, J. D. (1928–) U.S. molecular biologist, 139, 142

Wheeler, W. M. (1865–1937) U.S. zoologist, 66

Whitman, C. O. (1842–1910) U.S. zoologist, 66

Wilkins, M. H. F. (1916–) British biophysicist, 139

Wollaston, W. H. (1766–1828) English chemist and natural philosopher, 204

Wollman, E. L. (1917–) French molecular biologist, 137–46